Inventions,

from an Idea to a Business.

Arif Al-Awadhi

First Edition 2022

Printed in the United States of America.

Library of Congress Control Number: 2022946743

ISBN	Paperback	978-1-68536-848-7
	eBook	978-1-68536-849-4

Westwood Books Publishing LLC
Atlanta Financial Center
3343 Peachtree Rd NE Ste 145-725
Atlanta, GA 30326

www.westwoodbookspublishing.com

And that there is not for man except that [good] for which he strives (39) And that his effort is going to be seen (40)

Verses from the Quran chapter of An-Najm.

Index

Introduction

After publishing the second book titled "How to Become an Inventor?", I decided to release a second edition for it with additions. However, the owner of Dar Al-Hekma Library, Abdullah Sayyadi, advised me to write a new book instead. He said it would be a better option and it would bring more interest than releasing a second edition of the previous book.

The information I was planning to add to the second edition of the aforementioned book was not enough to form an independent book. I had to extensively read and research the internet, books and any other possible means for information that was relevant so that I may write a new work with new and useful information. This book has taken an arduous amount of time and effort. A total of three years.

As with my approach to my previous two works, I tried to make certain that the information within this book is simple and easy to digest. I avoided delving into complicated scientific theories that only specialists may comprehend. I want this book to be accessible to readers of different academic backgrounds and levels.

This book covers several aspects: Innovation and brainstorming, technical aspect, engineering aspect, administration and marketing aspect. All of these topics are briefly explored. You will notice throughout the book that the headings for chapters, and even sub-headings, could each be a title for a separate book.

I firmly believe in the importance of this book because I believe that inventions play a vital role in economic, social and even political developments in communities. Countries such as Japan and Korea do not have a natural deposit of petrol and gas and yet they have been able to achieve great things in their respective economies. This is through relying on innovation and industrial development, as well as their ability to turn innovation into commercial products that make their way into international markets.

Lastly, I wish for my efforts to be found useful by as many creative minds in our Arab countries as possible, especially those of youth, because they are our hope for the future. I hope that the reader will notify me of any observations about this work so that I may be able to develop it further in the future.

Chapter 1
Prelude

Inventions are a form of building human life and prospering with it on Earth. It is one of the things Allah commanded humans to do after creating them and land for them to stay on. Allah says in one of His verses:

((وَإِلَىٰ ثَمُودَ أَخَاهُمْ صَالِحًا ۚ قَالَ يَا قَوْمِ اعْبُدُوا اللَّهَ مَا لَكُمْ مِنْ إِلَٰهٍ غَيْرُهُ ۖ هُوَ أَنْشَأَكُمْ مِنَ الْأَرْضِ وَاسْتَعْمَرَكُمْ
فِيهَا فَاسْتَغْفِرُوهُ ثُمَّ تُوبُوا إِلَيْهِ ۚ إِنَّ رَبِّي قَرِيبٌ مُجِيبٌ ﴿61﴾))

<<And to Thamûd (people We sent) their brother Sâlih. He said: "O my people! Worship Allâh: you have no other ilâh (god) but Him. He brought you forth from the earth and settled you therein, then ask forgiveness of Him and turn to Him in repentance. Certainly, my Lord is Near (to all by His Knowledge), Responsive.>> from the Quranic chapter of Huud, Verse 61.

Defining an Invention

We can define an invention as a process of creating unprecedented products whether they are tangible such as medicine or machine, or intangible such as instructions to perform a task. A few examples would be the following:
1 – Enhancing the performance of an existing product, such as making a hammer or wrench handle more comfortable to wield.
2 – A new way to produce. Such as when Henry Ford created the first car assembly line.
3 – New breakthroughs for existing products such as the internet which was first only used for military communications.
As for the process of turning a new product or enhancer into a usable commodity, they are distributed and sold in stores and this is called 'trade'.
There is another synonym for inventions and that is creations, according to Arabic dictionaries. Scientifically, however, the two words have different implications. Inventions are, as previously mentioned, is the process of creating something never seen before. Creations, however, is to create something new and useful, or to use an invention in a more efficient manner than previously used. There may be other definitions that are similar. For this book, however, we are going to consider the two words to be the same.

Footnote: Our scholar, Muslim ancestors have always known that discovery, innovation and enhancing the human life are a way of worshipping Allah.
Inventions existed since Allah brought humans on Earth and they will continue to exist until the end of time. Inventions and discoveries are linked together and go hand in hand with one another. Humans long ago used to live in caves, eat raw food until they discovered cooking with fire. They learned making spears to hunt, using stones to create different tools such as hammers and blades, as well as piling up

stones to create a shelter. If not for inventions by creative minds through history, humans would still be living primitively in caves.
Inventions have no end, they are constant. Humans constantly go through change and development. Necessities are the basis of inventions, and an inventor is a creative mind whether they are young or old, woman or man, educated or otherwise. There is no doubt that Allah is the first and most creative inventor. He created different materials out of nothing, such as oxygen, nitrogen and all sorts of chemical elements, as well as solid objects such as iron, copper, wood and much more. We, humans, make use of those materials to create new things through mixing and shaping those materials that Allah had made in different ways and in different quantities.

The Goal of Inventions

Inventions are aimed to achieve the following essential aspects:
1 – Saving time. Cutting down on time leads to more production. For example, a car factory that relies on mechanical arms to build up cars could produce more cars compared to a factory that relies on human resources in the same time span.
2 – Reducing efforts. Compare an electric screwdriver with a regular one. Which of the two reduces effort during work?
3 – Saving money. The reduction of time and effort means more produce, and the conclusion of that is it leads to more profit.

Footnote: Have you ever used an electric screwdriver? Did you feel a big difference between that and a regular one?

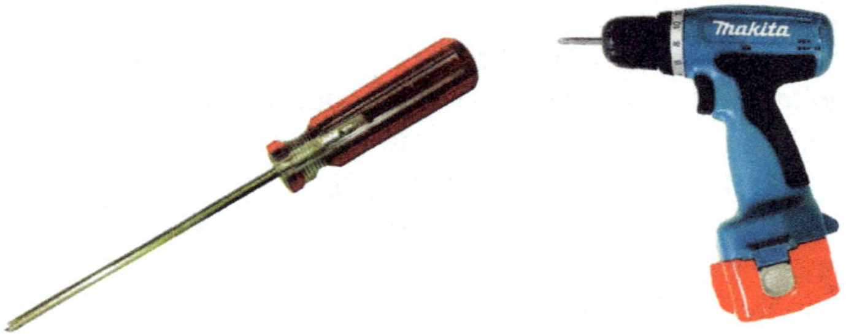

An electric screwdriver and a regular one.

Traits of Inventor

An inventor does not have to be an exceedingly intelligent person, nor do they have to be a person of high academic degrees. There are many inventions can be made with simple expertise, such as culinary tools and furniture. Many of these simple inventions have went on to achieve great success, and their inventors made millions. For example, correction pens were invented by a 17-year-old student who had not even fully gone through high school. This new invention gained a net worth of millions. In 1979, the product was sold to a company for 47.5 million dollars. We should not forget that this product was essential in a time where computers were not available yet. Every document was either hand-written or done by a typewriter. Some of the important traits that an inventor should have are observation, diligence and curiosity, always seeking answers, and to be able to turn an idea into a reality. In addition to that, general know-how and commitment to know all that is new with technology. These are some prominent traits usually found in inventors.

Footnote: A successful invention is a suitable solution at a suitable time.

The Path of Invention

Every invention started from a concept that went through several stages before turning into a new product that finds its way onto store shelves. This process may not be easy and may face several obstacles. Therefore, it is important for every inventor to be aware of different aspects related to this topic. Any given invention usually goes through the following stages:

1 – Determining the idea
2 – Feasibility study (Whether it is a practical product to use, etc.)
3 – Technical design
4 – Registering the patent
5 – Developing and creating a model
6 – Manufacturing and marketing

The sequence of the aforementioned stages does not necessarily have to be fixed, meaning not all inventions go through that same order. For example, the stage of registering the patent may be the second step after determining the idea, or in some cases, there may be no need for that at all. We will discuss this further in the chapter about patents. Marketing could come at the very end after manufacturing and so on. For the record, an invention can flop at any of the previously mentioned stages for different reasons.

We will begin to explore in-depth each stage as we go through the book, with an emphasis on the practical aspects without delving too much into theories or anything too technical that an average person may not understand.

Chapter 2
The Source of Invention Ideas

A strange yet wondrous idea. Its creator is truly a genius. I would like to peer into your mind and extract all its genius ideas.
This was a comment made by a student. Their friend at school had proposed a scientific concept regarding a particular topic. Indeed, that friend had been talented, a lover of science and inventions, and was the one that stood out amongst their friends.
However, is bringing a creative idea the only thing a gifted inventor is capable of? Where do these ideas come from? This is what will be in this chapter.

How are ideas created?

Ideas for inventions are born from the interactions between human beings and the environment around them. Using different tools throughout the day generates ideas. This is because while using these tools, a person will think about how to enhance them for better efficiency, or a person will feel the need to come up with something that will aid their work and make it more seamless. This is precisely what the well-known saying, "Necessity is the mother of invention", refers to.
A necessity can be defined as a shortage of something with a human being. It could be subtle. Therefore, it is necessary to determine whether there is one, and once that is done, one can proceed to find a suitable invention which will in turn sate this necessity.
Fear also has a role in coming up with inventions, but how is such a thing possible? Experiencing fear most likely happens because of the existence of an undesirable cause. This fear pushes us to think in order to find solutions that help us avoid the that cause, or to reduce its impact. For example, people are afraid of running out of fossil fuel as an energy source. This incentivized them to find alternate sources of energy such as solar energy, wind energy and others in order to ensure that life can go on.
People's fear of natural disasters also pushed them into searching for solutions to either stop them from happening or to reduce their impact as much as humanly possible. An example of this is the development of early warning systems for earthquakes, as well as enhancing systems of fire detection and extinguishing in different kinds of buildings.

Regardless of the different incentives one might get, creative ideas stem from our bright minds that have been gifted to humans by Allah. Not all minds are the one and same. There are those that quickly realizes the existence of a problem and the need to innovate something and find solutions to solve it, and there are others that go through these problems and do not think or even realize the need to develop.

To be a creative inventor, you must think outside of the box. Let your mind explore the vast world of possibilities and search for ways to develop the idea of invention. Be flexible, be practical and be far away from complications. Keep away from all that distracts your mind and keeps you from focusing on the invention. The responsibilities of life and all that entails with them will interfere with the clarity of your mind and with proper useful thinking. We are unfortunately unable to completely detach ourselves from the problems of society for as long as we live, but we can try. Do not hesitate to jot down any idea that crosses your mind no matter how trivial it may seem.

In exchange, not all inventions are based on the idea of solving certain problems faced by a human being. There are inventions that were made without the existence of a problem that called for it, but by mere human curiosity. For example, the invention of spaceships happened due to the curiosity of human beings and their love for exploring outer space, other planets and to find out more about this vast universe that we are a part of.

We must not forget to acknowledge that not all ideas are applicable. There are thousands of ideas that may come up but only a small fraction may turn into reality. There are many reasons for this, such as high production costs, marketing, the practicality of the product, its effect on the environment, etc. Any given idea has no value if it does not have the potential to be turned into a real and practical product that we can benefit from.

Figure (2-1) from the Canadian Innovation Centre shows us the vast number of ideas. They go through seven stages of refining only to end with one idea remaining, and this is according to proven statistics from the Canadian Innovation Centre. The chart also shows that out of 3000 raw ideas, only 3% are presented. 125 of those ideas are turned into small projects that go into the stage of effective development. They then come down to four and then one. This one idea is turned into a successful commercial product that people benefit from.

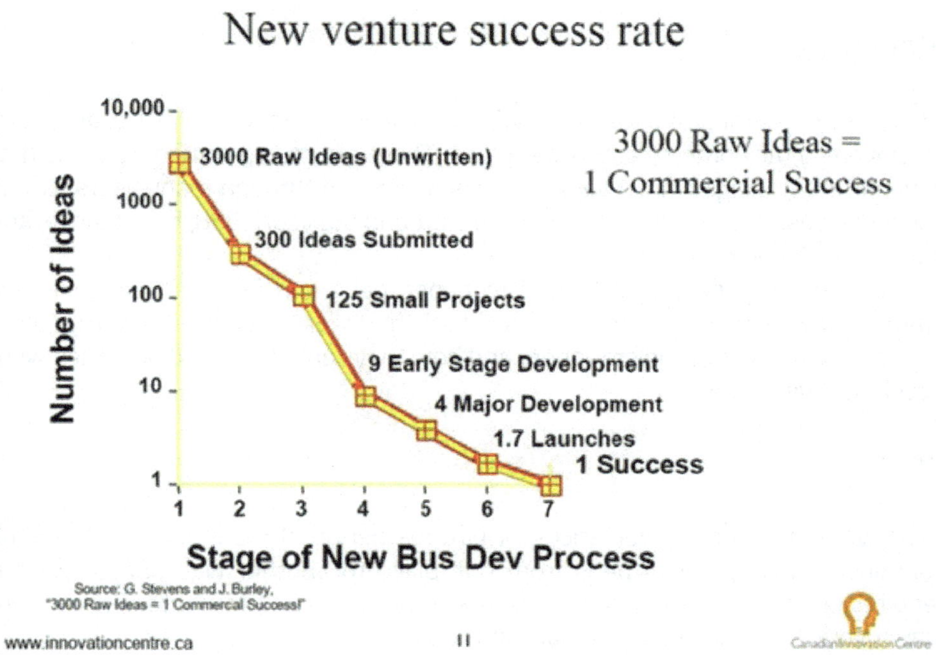

Figure (2-1) A chart showing the success rate of ideas

One possible reason for the failure of an invention is the lack of a sufficient study for the concept, and therefore a lack of understanding for the potential problems and how to avoid them, whether it is from a design perspective or a marketing perspective, as well as the suitability of the product for new users, among other reasons.
Therefore, we will explore how to study the feasibility of a new invention in the third chapter before we dive into the application of the idea; to save on time, effort and funds.

Look around you

Look around you at home, at your workplace, at your school, at your club, or any place you go to. Look at the devices and tools. Then ask yourself: How were those invented? Let us look around us together and ask ourselves this question, through some examples, and try to see and learn what may have led up to the invention of those creations.

Stationery

We all use different stationery tools whether we are at work in an office, at school, at university, at home or any other place. There are many stationery tools that have different usages, and every tool was developed through many years. Let us ask ourselves: How did stationery look like around the year 1888, for instance, and what do we have now?

In the past, stationery was limited to paper, pens and other simple tools like stamps. Now, however, there are tens of tools that were not available at that time. Let us look at some of the stationery and how they came to be and how they were developed with time.

Pens

Pens were made of wooden sticks, or bird feathers that were dipped into an ink container. Carrying these pens from one place to another was difficult, and it required using a special bag to put these wooden sticks and ink containers as there was a very probable chance of spilling the ink.

Around the year 953, the fourth Fatimid Caliph, Al-Mu'izz li-Din Allah, created a pen that did not smear his hands or his clothes. Based on that, a new kind of pen was made that had a tube containing the ink, and a tiny opening on the tip of the pen where the ink was able to pass through while writing. When not using the pen, it would be put upside down to avoid any ink coming out. This was the fountain pen ever made. (Source: https://en.wikipedia.org/wiki/Fountain_pen)

Note how the idea for a new kind of pen stemmed from its user's realization of the flaws of the traditional pens of the time, and the need for a new kind of pen that performs better.

In the next decades, others would make similar kinds of pens with slight changes to enhance its performance further, and this kind of pen is still used to this day. Look at the figure (2-2).

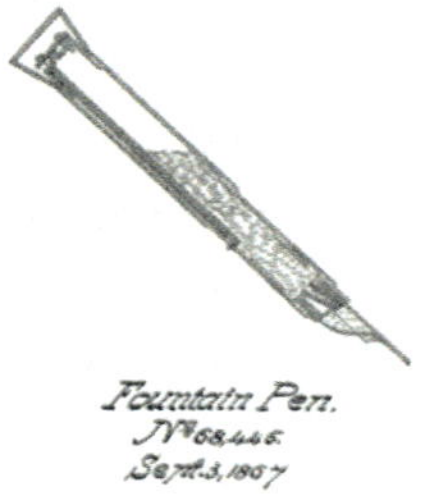

Figure (2-2) depicts a type of fountain pens that was created in 1867. (Source: Wikipedia)

In October of 1888, John Loud invented the ball-point pen that is characterized by a small, metallic ball shaped tip that spins while writing to transfer ink to the outside as shown in the Figure (2-3). Loud worked in the business of tanning leather and he found it difficult to use a fountain pen to write on rough surfaces like leather. (Source: https://en.wikipedia.org/wiki/Ballpoint_pen)

Figure (2-3) A ball-point pen's metallic ball-shaped tip. (Source: www.daysoftheyear.com)

The inability of fountain pens to fulfill what was needed, writing on leather, led John Loud into thinking and thus inventing yet another type of pen. This newfound type was further developed in the coming decades.

Paperclips

In 1867, Samuel B. Fay invented the first paperclip and it was a metallic wire folded over itself as shown in the figure (2-4). Fay originally created a paperclip to attach tickets to fabric, as using glue or pins would damage the fabric. It was later used to hold papers together. (Source: www.ideafinder.com/history/inventions/paperclip.htm - https://www.officemuseum.com/paper_clips.htm)

Once again, the inventor, Samuel B. Fay, realized the need for such a tool, based possibly on his work. In the coming years, different kinds of paperclips were made.

Figure (2-4) Samuel Fay's Paperclip
(Source: www.officemuseum.com)

Figure (2-5) Modern paperclips

Staplers

During the 19th century, the increasing use of papers required inventing a practical way to hold linked documents and papers together. Before the staplers that we know today was invented, papers were bound together using a thread or a pair of nails.
In 1854, Hymen L. Lipman enhanced the stapler machine. It was a big machine that would be placed on a table as shown in the figure (2-6), and it had its flaws.
(Source: http://www.officemuseum.com/staplers.htm)
This tool was enhanced over years and decades until it reached the point that we know of today.

Footnote: Try to study all types of staplers available in the market and think of ways to develop it further.

The need for staplers happened because of the increasing use of papers, and it is one of the things that led to the prospering of office work, most notably, documentation.

Figure (2-6) Hyman L. Lipman's stapler **Figure (2-7) A modern stapler**
(source: www.officemuseum.com)

Staple Removers

Footnote: You use this tool every day. Have you noticed any flaw in it? Have you thought about how you can enhance it?
Staples would sometimes need to be removed after some time. This led to the need to invent yet another tool. This tool came into existence in the form we know in 1932 by William G. Pankonin. (Source: https://en.wikipedia.org/wiki/Staple_remover)
Thus, an invention leads to another invention.

Figure (2-8) Staple remover

Folders

With the increase in paper usage and having different categories of documents came the need of holding papers of one category together to protect them from getting lost or being damaged. This would also make categorizing much easier, and so folders were invented by Friedrich Soennecken in 1886 in the German city of Bonn. (Source: https://en.wikipedia.org/wiki/Ring_binder)

Figure (2-9) Folders

The creation of folders led to an immense amount of convenience with office work. In addition to keeping documents together, it helped save time by categorizing documents in an orderly manner. For example, electricity and phone bills were now put together in one folder, papers related to sales in second, papers related to purchases in a third folder, and so on. The contents of a folder would also be written on the side, and they would all be put together side by side on a shelf. Therefore, evolution of life makes us think of new, creative ways to facilitate work performance.

Paper Punchers

As a result of folders coming into being, paper punchers were invented to pierce holes into papers to hold them together in folders. Matthias Theel invented the first paper puncher in the late 19th century in Germany. Friedrich Soennecken then went on to develop a new type of paper punchers in 1886. (Source: https://en.wikipedia.org/wiki/Hole_punch#cite_note-1)

Figure (2-10) Soennecken's paper puncher from 1888 (Source: Wikipedia)

Footnote: Papers created the need for folders. Folders created the need for paper punchers. Will paper punchers create the need for a subsequent invention?

Photocopiers

Footnote: Was inventing a photocopier easy? What kind of skills did it require? Was Chester alone in thinking about an invention like this?

Copying papers in the past was done by hand. However, re-writing was an extremely arduous task, and the possibility of making a mistake while copying was very likely. In 1937, Chester Carlson invented the first photocopy machine.

Chester was born in Washington on the 8th of February in 1908, and he was the only son of his father that worked as a wandering barber. His mother passed away when he was seventeen. After graduating school, he went on to attend a college, specializing in chemistry, and then went on to study in California Institute of Technology. He graduated after two years with a specialization in physics.

After graduating, Chester struggled to find work. He eventually found work as an engineer in Bell Telephone Laboratories. Later, he moved onto another job at a law firm and this would be a stable job for him. This new job, however, required him to study law, and he did so during nighttime.

With the nature of this job, Chester was dealing with big amounts of documents that needed hand-copying, and this was a tedious and tiresome task, especially with the need of proofreading to make sure no mistake was done while copying. Because of the nature of Chester's job and environment, he came up with the idea to make a photocopy machine to save time and effort. Chester was certain that such an invention would make a significant difference. (Source: www.ideafinder.com/history/inventors/carlson.htm)

Figure (2-11) Chester with his new machine (source: news.xerox.com)

Opening Cans

Footnote: Was Fraze the only person in the world to fall into a dilemma like this one? Were there others who thought about a solution for it?

Pop-top cans have a unique and fun story. Ermal Fraze (1913-1989) was out on a trip with his family, holding what he needs to have an enjoyable time. He suddenly realized that he forgot to bring the can opener. This made him make use of a sharp, metallic part of his car frame to open the can. For months after this, Fraze was often contemplating about that situation he was in, trying to find an easier way to find metal cans without the need of another tool. He found what he was looking for in 1959, with a metallic can that had a part you could pull on at the top to open the can. This invention is still used to this day as seen in figure (2-12).

Figure (2-12) a round grip to open a can

Fraze's realization for the problem and its need for a solution, in addition to his diligence and efforts led to this invention.

Fraze founded a factory to produce and sell this new kind of self-opening cans in the 60s of the past century. By 1980, Fraze's factory's profits reached 500 million dollars. (Source: http://web.mit.edu/invent/iow/fraze.html)

There are fields that are ripe with opportunities to generate new ideas. An example is the field of tending to the disabled. For instance, A person missing one finger needs a solution different from a person missing two fingers, and a person missing two fingers needs a solution different from a person missing all of their fingers, and a person missing all of their fingers needs a solution different from a person missing a hand, and a person missing a hand needs solution different from a person missing two hands, and a person missing two hands needs a solution different from a person missing an arm, and a person missing more than one limb needs a different solution as well. You could put an infinite number of disabilities that need different solutions, and on top of that, the daily circumstances of a disabled person are what determines the solution. For example, a person missing four fingers that wants to use a computer needs a solution that is different from someone missing four fingers that wants to cook, do carpentry, etc.

Let Us Think

After looking at how different ideas are created, let us try to think of some scenarios and try to discover any shortages and brainstorm.

Footnote: Have you read about the history of the invention of bicycles? Who was the first inventor? How did it look like? How did they evolve through time until now?

Think about the environment that you live in.
What is there around you? What are the tools that you use?
For example, do you use a bicycle?
Have you ever felt that your bicycle is missing something? Or have you had a problem while riding it?
Did you ever think that if you were to change something with it, it would perform better?

Figure (2-13) What ideas do you have to improve this bike?

You share a bicycle with your friend or little brother, and because of the difference in height between the two of you, you must adjust the seat height every time by taking off the seat screws and re-tightening them. It is a simple process, but it will get boring with time. How can we do this process in a fast and efficient way?

Another example, are you a chef?
What tools do you make use of in the kitchen?
Are there tools that you use that could be enhanced for better performance?
Do you feel like you need to create new tools to help you save time and effort?
Have you ever thought of creating a new food product? There are plenty of food products that did not exist in the past and then when they did come to existence, their inventors made millions out of them, such as ketchup, mayonnaise, and other sorts of sauces. This extends to drinks as well such as Vimto, rose syrup and many other examples.

Figure (2-14) different food products

A Brainstorming Session

Brainstorming sessions are an effective method in generating ideas and incentivizing the mind, as well as exchanging experiences between people in a project in order to find creative solutions to a problem.
But in order to benefit from a session as much as possibly, there must be a few rules to adhere to, and they are the following:

The first rule:

Determining the objective of the session. The objective is the problem that we need to generate creative ideas to solve. For example, we want to come up with ideas to enhance the performance of bicycles, that would be the main objective.
Then, a series of questions would be put into place. For instance: Can we increase the speed of the bike? How can a bicycle be safer? And so on.
In addition to the main objective, there are secondary objectives. What if we wanted to provide the bike with an air conditioning system that can generate enough electricity to power that system?

The second rule:

The session consists of two stages. The first, is to think of as many ideas as possible without thinking too much about the downsides and without criticism no matter how strange, trivial or funny an idea may be. The second stage is to evaluate all the ideas and to accept the most practical and potent ones.

The third rule:

The more people involved with a passion or any sort of prior experience relevant to the project, the easier it is to come up with ideas. For example, we cannot ask someone who has never worked with car engines to think of ideas on how to enhance an engine!

The What-if Way

It is also possible to throw around ideas by starting a question with "What if it was like this and that?" or "What if we did this and that?"
Right after that, we try to see the feasibility of the proposal. For example:
"What if we made a bike with four wheels? What is the point of that?"
"What if we added a fifth wheel to cars? How does that extra wheel bring any benefits?"
"What if we made a car with only two wheels? How can we do that?"

Standard Procedure in Companies

Manufacturing companies follow a particular system in the creation of ideas for new products, or to upgrade and enhance existing products, and that is through communication with clients. The stages of this system can be explained in the following diagram:

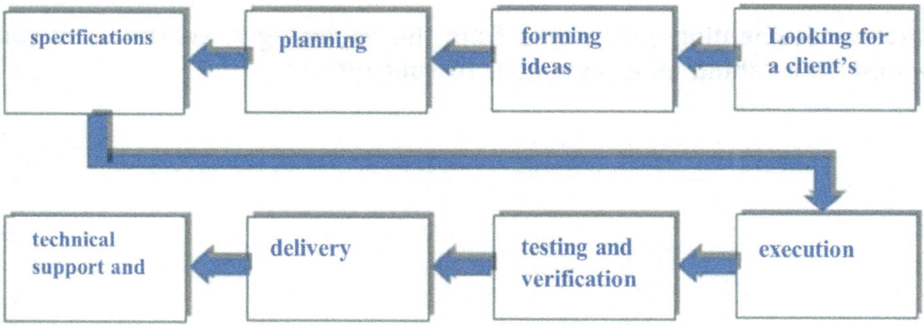

Company agents collect information from clients through available means, such as field visits, interviews aimed at knowing their day-to-day tasks and to study their needs that facilitate their work. From this, they get new ideas to work with, analyze and study in more than one way, including market competition and choosing the best.

A conceptual framework for the new product is then made, listing its technical specifications, so that it can move onto the execution stage in a lab and test the performance and efficiency through prototypes.

Invention Notebook

In order not to forget about your own ideas for inventions, make sure to constantly write them down in a notebook or an organized folder with the following information mentioned for each idea:

- Date of idea creation
- Idea name
- Idea explanation
- Source of the idea
- A diagram for the idea
- Required parts and materials
- Titles of any relevant book or internet website references that could help you carry out the project

Footnote: Documenting your works from the beginning is essential, as your invention notes stand as a testament to your efforts.

Conclusion

As mentioned in the start of this chapter, necessity is what drives us and pushes us into inventing new things, and multiple inventions can follow after one. Fountain pens came as a result of primitive pens, ball-point pens came as a result of fountain pens' inability to write on leather. Paperclips, staplers, folders and paper punchers all are inventions that came into being because of the rising usage of papers, and thus the need to organize and maintain them.

Every inventor would not have made their creations if not for the problems that they faced that made them realize the need to innovate and come up with new tools to help them with those problems.

Coexistence and interaction with the surrounding environment are what lead humans to think.

I realize that there are thousands that share the same environment and circumstances as you, but you may be the only one to grasp the idea of inventing something new. That is because you are creative, and you always think of changing for the better. Always remember the saying "There are plenty of fish in the sea, but the one with the hook is the one that be caught". That is why, if you get any idea, make sure first and foremost that it is new, and then write it down to ensure ownership rights, and then work on turning it into a reality before others beat you to it.

Chapter 3
Studying Project Feasibility

As previously mentioned in Chapter 2, not all invention ideas are useful. Very few of them are turned into useful creations. Which is why, before attempting to work on an invention, we must first conduct a study of feasibility, in which the feasibility of any given project is thoroughly researched, with all of its varying advantages and disadvantages to society. This is important in order to evaluate the estimated amount of interest in the product and therefore the estimated profit.
It is also imperative to seek out any other similar inventions, as this will save you precious time and effort by not having you repeat the same experiments others may have already done. In addition to that, it will help you learn information that may be insightful and applicable to your own invention.
We will discuss how to look for similar inventions and to search for all the necessary information at the end of this chapter.

Feasibility Study Aspects

Feasibility study can be divided into two parts:
1 – The practicality of the invention. In order to be certain of whether an invention concept can be turned into reality, we must ask the following questions:

- Are the raw materials required available?
- Are the necessary tools to create the invention, such as computers or machines, available?
- Are there any sorts of pre-requisite experiences required?
- Will the invention truly work as intended?

2 – The extent of marketing the newfound invention after its creation. This, in addition to the following questions will help us conclude the feasibility of the invention:

- Who are the users of invention?
- How is this invention important to its consumer?
- How much are the production costs?
- What are the side effects of the invention?
- Does it conflict with the culture of a society?
- Does it conflict with local regulations?
- How easy is it to use the invention?
- How long is the invention expected to last?

These questions are criterion to keep in mind to measure the feasibility of an invention and how well it could be converted into a commercial product. We will discuss each question and then compare them with each other and how to apply them.

Naturally, if a feasibility study fails to go through the first part, then there is no need to take it to the second part. That will only be a waste of time and effort for an invention that will likely not convert well into an invention.

i. The Invention Users (number of users)

Who does the invention seek to pander to? How many of them are there?
The more people it panders to, the greater the chances that it will be a commercial success, and therefore a successful invention. There are inventions that were made for every individual part of a society, such as cars, lights, all kinds of temperature machines used for either cooling or heating, etc. There are other inventions that were made for more specific users, so an invention made to aid the disabled will be limited to them, and if the number of users is small, then the invention will have limited sales, and vice versa.
We must not forget to point to the necessity of creating the right invention at the right time, as high demand will only serve to increase the value of that invention. There are inventions that flopped because of going against this.

ii. The Invention's Importance to Consumer

What makes the invention important to its consumer? Will it be of great or little importance?
Is the invention a luxury or a necessity that fills a gap of need in a person's life?
For example: Is it important to the consumer for a phone to have multiple applications, or is it sufficient to have a phone that only sends and receives calls and messages?
Another example: If the invention is aimed at a specific user of disabled people, will it help them immensely, or will it be of limited help?

Naturally, the more convenient an invention is, the more likely it will be acquired by more people. Therefore, it is highly advisable to conduct surveys aimed at the target users in order to determine their need for it. This may require manufacturing some prototypes to run tests, as it would otherwise be difficult for a consumer to evaluate the invention.

The need for any given invention could be temporary. For example, during natural disasters, people would require certain emergency tools or ways to help accommodate those who lost their homes. When Japan had a series of earthquakes in March of 2011, the impact damaged Fukushima's nuclear plants which led to the shortage of electricity. A Japanese based company called "Green House" developed a type of light known as light-emitting diode (LED) that was powered by water and salt. This new invention did not require any batteries. It consisted of two bars of magnesium oxide soaked in a bag of 350 milliliter of water mixed with 16 grams of salt. This would be creating light that would last for eight hours.

Footnote: Picture a village in a remote region with mountains, and installing electrical currents was not possible. Will green energy ever thrive?

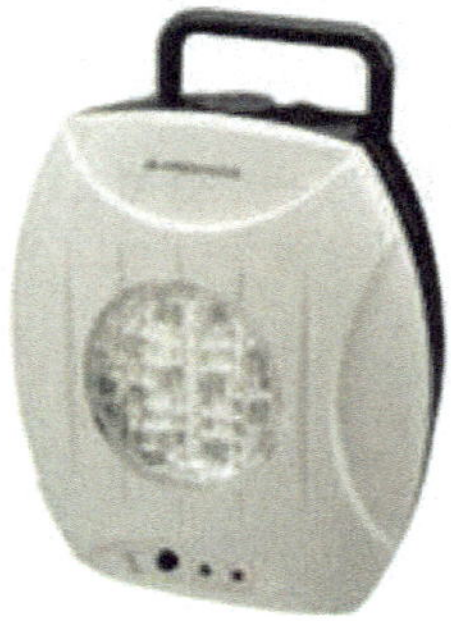

Figure (3-1) A light that is powered by water and salt.
Source: (http://www.aljazeera.net/news/pages/f3f77454-43b5-4a29-8c47-f8d7456da856**)**

A tool such as this making use of accessible materials is truly efficient in emergencies of that nature.

iii. Production Costs

What are the production costs to create this invention? Will it be expensive or cheap? Will it be possible to ship it out with a suitable price tag for those that need it?

Without a doubt, the less it costs to produce the invention, the easier it will be to sell it, and the higher the costs are, the more expensive it will have to be sold for in the markets.

Consumers always compare their need for any product in a market to the price tag. We sometimes resort to buying expensive products because of a dire need for them, like with medicine.

Footnote: Energy-saving lights are six to seven times the price of regular ones. So why is it that many people choose to buy the former?

We must also understand that the economic circumstances of a country have a hand in the sales of inventions that are considered luxuries. The stronger an economy is, the higher the sales and vice versa. If an economy is in a rough state, people will most likely latch onto necessities only.

That is precisely why an inventor should compare the demand of society and the invention, and to think about the lowest price the product could go for, in order to determine whether it will sell well.

An invention may be high in price due to expensive materials required for its production, in addition to other factors such as shipping prices, etc.

Figure (3-2) shows a comparison between demand and pricing.

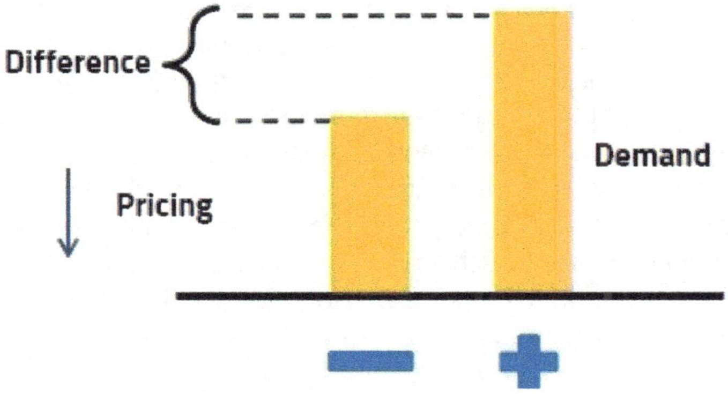

Figure (3-2) Comparing demand and pricing

The higher the demand for a product is, the more likely it is for an invention to perform well commercially. Try to think of an example with any product in a market. You, as a consumer, decide whether the price tag for a product is reasonable based on your need for it. This goes to show that your invention will not sell if the price tag is higher than the demand.

In turn, however, if there was a notable demand on a product, and the quantity of it was low and does not meet the demand for it, there may be room to increase its price because it is a desired product. Manufacturing a new invention in ample quantities helps reduce production costs and therefore sell the product for a lesser price.

An example of the difference in demand in pricing is when some companies or stores make tempting offers that tend to be limited to, say, one day, on a certain product. This offer would entail reducing the already-reasonable price of the product. This can cause extreme levels of crowdedness in these shops, as people want to procure the product at that discount before it runs out of stock. In some instances, you would even have people waiting in line outside of a shop from the early hours of the morning.

iv. Invention Side Effects

Are there any side effects to this invention?

Side effects may be limited to the consumer, or they may extend to the environment around the consumer, whether it is other people, animals, plants, etc.

There are inventions that achieved great success, such as different kinds of medication, but their side effects started appearing long ago. For example, DDT is a pesticide chemical compound that was created by Austrian chemist Othmar Zeidler in 1874. World War 2 (1939 – 1945) caused a lot of damage and destruction to European infrastructure, and because of disease-spreading insects, a lot of diseases such as malaria were spreading throughout the continent. This caused Swiss chemist Paul Müller to develop this chemical compound and his efforts would have a significant impact in ridding Europe of a grave pandemic. As years went by and in the 1950s, researchers found side effects to this pesticide that harmed the environment because it did not decompose over time. This was a factor in the harm of beneficial insects such as bees and butterflies, and it also caused harm to fish because the pesticide found its way into rivers. The pesticide was banned after official decisions were made in the late 70s. (Source: Wikipedia)

v. Conflicting with a Society's Culture

Can the culture of a society have an impact in preventing the success of an invention? Will there be a need to spend more time, money and effort to try and convince the society to accept the invention?
Or is it that this invention is simply incompatible with some societies?
For instance, an invention may be a food product that contains alcohol or substances from an animal like a pig. Could an invention like this see the light of day in a Muslim country?
Of course, because of the religious prohibition of eating pork and its derivates, an invention like this will never succeed in Muslim societies.
A world-class fast-food chain opened a branch in India, and with time, its executives realized that there is not much demand on their food unlike their other branches in other countries. They later discovered that the residents of that Indian area were vegetarians and did not eat beef or chicken.
Therefore, it is vital to ascertain whether an invention conflicts with the culture of a society, even if it seems unlikely to be the case.

vi. Conflicting with Local Regulations

Local rules can be an obstacle to some inventions. For example, the governments of some countries prevented the entry of satellite channel receivers for religious or political reasons. They also blocked many websites on the internet for the same reasons.

vii. Ease of Use

Is the invention practical to use?
In other words, will its user have trouble trying to use this invention, or is it simple enough to use that there will not be much trouble at all?
Any invention could be useful, but at the same time, that does not necessarily mean that it will be a practical invention. Perhaps its lifespan is short. Therefore, in order to get rid of its shortcomings, many inventors seek to enhance the invention as much as possible, and this is a process that may take long years.

viii. Invention Longevity

Is the invention a disposable product? Or perhaps it has disposable parts?
Disposable parts play a prominent role in establishing a long-term relationship between the consumer and the manufacturing company. The company sells the product once to the consumer and then continues to sell its parts for many years. We have many contraptions that have disposable parts. For instance, desalination devices contain a filter that is replaced every few months. Printers run out of ink and that raises the need to go and buy more. Cars are also another example, etc. If a product had disposable parts, this becomes an advantage and contributes to additional revenue for the invention.
On the other hand, it could also backfire. The constant need for switching parts may become tedious for the consumer, or expensive, and that may turn them reluctant to continue with it.

Footnote: Some companies assemble their devices for consumers for free, and they would also offer free routine maintenance, and in turn, they for satisfy customers and ensure they continue to use their products and services.

Now let us compare the eight points of criteria. We will divide them into two groups. One positive and one negative. The positive group contains criteria that must be nurtured for a more successful invention, while the negative group contains criteria that should be reduced or removed. Look at the table below.

Positive Group (+)	**Number of Users**	**Demand**	**Ease of Use**	**Longevity**
Negative Group (-)	**Side Effects**	**Regulations Conflict**	**Cultural Conflict**	**Pricing**

Figure (3-3) comparison table

Based on this division, we will have a scale that in the shape of a seesaw. We will measure through it the success of an invention. Look at figure (3-4)

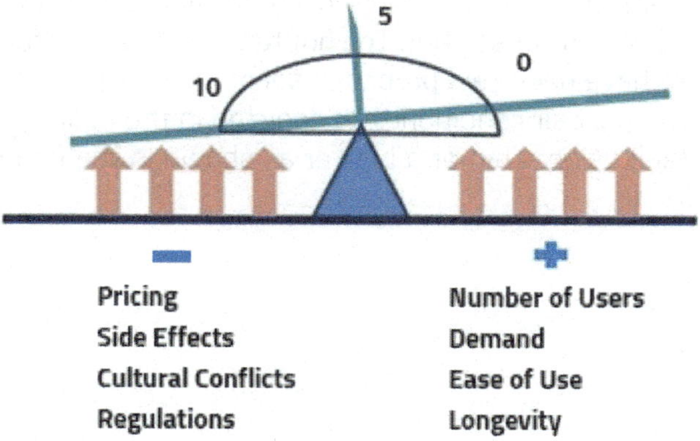

Figure (3-4) Invention Success Scale

Looking at this figure, we can explain the process of invention evaluation and feasibility study. The figure shows the positive group on the right and the negative group on the left. The stronger group will push the other side of the seesaw upwards. If the positive group outweighs the negative group, then the right side of the seesaw will go up and we would have a high chance of success. If the negative group outweighs the positive group, then the left side of the seesaw will go up and that would indicate a low chance of success.

If one of the positive criteria does well, this will give score to the positive group and take them higher, and the same goes for the negative group. If one of the negative criteria sees an increase, it would take the negative group higher. For instance, extreme demand for an invention would indicate a high chance of success, even if it caters to limited users and the price is high as with some types of medication. Accordingly, an invention may see failure if it is not compatible with the culture of a society or goes against their local regulations, even if it has many consumers and there is demand for it.

Another example could be with a medication for an epidemic disease like malaria or tuberculosis or anything else. This is a product of utmost importance. Let us picture that a disease spreads in a poor country that is unable to afford the costs. A high price would lead to the failure of selling the medicine, even if the demand was in great numbers.

Now let us attempt to apply these criteria on some inventions to evaluate their potential.

Water Tank Coolers

Constantly having hot water in the Middle East has been a problem for decades, especially during the summer, where the temperature can get close to 50 Celsius degrees. The root of this issue stems from desalination plants, as water would come out of the plants already hot. The hot temperature only further amplifies the problem. There have been two proposed solutions to this: Cooling a home water tank using a compact air conditioner as depicted in the figure (3-5), or to cool the home water tank using a fan or a blower as shown in the figure (3-6) to replace steam with dry air.

Figure (3-6) Cooling a water tank with a fan

Figure (3-5) Cooling a water tank with an air conditioner

	Fan	Air Conditioner
Number of Users	Very high	Very high
Demand	Very high	Very high
Ease of Use	Easy	Easy
Disposable Parts	Dust filter	Dust filter
Side Effects	Low	Low
Regulation Conflict	None	None
Culture Conflict	None	None
Price BD	40-50	60-80

Figure (3-7) comparison between invention success

Let us compare between the two solutions.
By referring to the previous table, we see that the two solutions share similarities in all their criteria except for the price. This is because they are two solutions to the same problem. Meaning there is competition between the two, and naturally, the cheapest option will have more sales.
In the aforementioned example, using an air conditioner to solve the problem is not only more expensive, but it also costs more to maintain in the long run. It uses more electricity, and it needs routine maintenance as with any other home air conditioner.

Compressed Disks (Zip Drive)

Compressed disks, which were popularized by Zip Drive, is a tool for data storage that was produced in the end of 1994 boasting a capacity of 100mb and later 750mb. It was priced at around 200 dollars and one disk was priced at 20 dollars. It was the perfect tool to move big files for those that needed it. Look at figure (3-8).

Figure (3-8) Compressed disk with a Zip Drive
(Source: http://goughlui.com)

Despite its high capacity at the time, its data transfer speed was slow compared to the lower-capacity floppy disks, which held a maximum of 1.44mb. It also had technical faults, such as incompatibility with some types of computers, which could bring down the transfer speed even more.

Sales for this product decreased between 1999 and 2003 after the appearance of compact disks (CD) that could hold 700mb of data in it. It was light, cheap and faster in data transfer. Some time later, USBs were invented, and they were a massive success in terms of price, weight, and data capacity. (Source: https://en.wikipedia.org/wiki/Zip_drive)

We can see that Zip Drive was a success at first because of its ability to store massive files. It did, however, have technical problems such as the aforementioned incompatibility and its slow speed which eventually rendered it impractical. Then, better-performing alternatives started appearing, and they were cheaper and more efficient for their time.

Any invention going into the market will garner reception based on the users and the demand for it, but product faults will naturally drive consumers away over time, because this gives competitors a chance to create better solutions that are more practical and thus attracting consumers to them. Figure (3-9) shows this process of product lifespan.

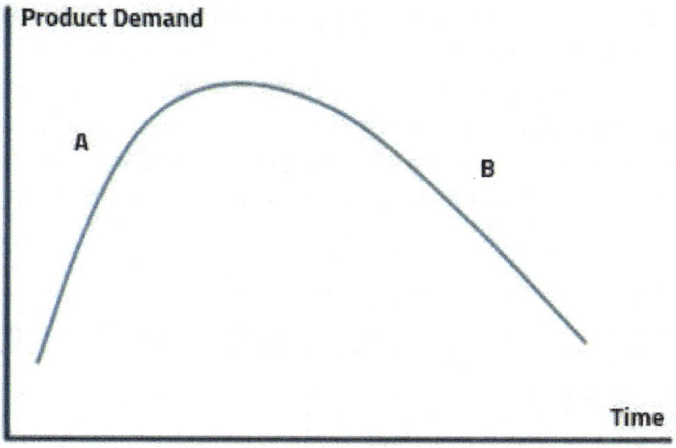

Figure (3-9) The rise and fall of a product

At first, interest for a product goes on a rise as with (A), and if it had faults that hinder its performance, there will be others that use this to their advantage to create alternative solutions based on the same idea of your creation. Leading to the fall of the product with (B). It is important to recognize the faults in your invention and work on fixing them before others exploit this situation to their advantage and create a better alternative.

Electric Car

Many of us have either read or heard, at least through the TV, about electric cars. Electric cars have an important advantage, and that is its eco-friendly nature. It relies on electric energy completely to turn on its engine and the other car devices. Unlike traditional cars that are powered by fuel, which produces many harmful gases, such as carbon monoxide, carbon dioxide and so on.
And now the question: Why, then, are fuel-powered cars still rampant?
To answer this question, we must compare the electric car to the traditional fuel-powered car based on the criteria we previously mentioned in figure (3-10).
Electric cars get their supply of power from rechargeable batteries, charged in their own special devices. There are many types of electric cars. We will now try to measure their criteria in the next table according to information from the U.S. Department of Energy.

	Traditional Cars	**Electric Cars**
Number of Users	Very high	Very high
Demand	Very high	Very high
Ease of Use	Easy	Hard
Disposable Parts	More	Less
Side Effects	More	Less
Regulation Conflict	None	None
Cultural Conflict	None	None
Price	Cheap	Expensive

Figure (3-10) comparison between electric and traditional cars

Footnote: Do you now understand why we do not use electric cars? Do you have reasons why they are not feasible?

As for ease of use, this is the one criterion that prevents electric cars from being an alternative to traditional cars, and here are the reasons:

Battery recharge: Recharging an electric car's battery to maximum takes about twenty to several hours depending on the size of the battery and the power of the charger. While traditional cars need a minute and a half to refill its gas tank.

Distance: A fully charged electric car battery can drive you about 160 kilometers only, and even then, it depends on driving conditions. Driving up hills, carrying heavy baggage, using air conditioner during the summer, etc. All of these take away more battery power, while traditional cars drive at least twice that distance when fully fueled.

Speed: The speed of electric cars does not quite match that of traditional cars, and the higher you go with speed, the more power is drained.

After looking at the things mentioned above, we can see that traditional cars are more practical to use, especially in countries where governments have fuel support, reducing the price of fuel for people.

There have been some solutions proposed to enhance the performance of electric cars, and these are:

1 – Providing car battery charging stations in every parking lot and home so that the driver is able to recharge the car more often. Like say from night to morning at home, and then from morning until after work in the parking lot.

2 – Devising an electricity generator that is powered by fuel to recharge the car's battery in the event that the car is stranded somewhere far away.

3 – Installing an engine that is powered by fuel in the event that the battery power runs out. Mixing a fuel-powered engine and an electric one would at least reduce the use of fuel.

So then, when would the usage of electric cars be suitable?
In golf courses to transport players, in football fields to escort wounded players, in big airports to take passengers into the airport building, as well as in internal streets like in universities to move from one building to another. All these places mentioned have short distances and do not need high speed.

Therefore, electric cars' technology is constantly evolving, and as we see, it is increasingly spreading around the world. The manufacturers of such cars spend millions on R&D to improve their efficiency and to be more suitable for people's needs.

Figure (3-11) An electric car recharging its battery
(Source: http://www.electricauto.org/resource/resmgr/docs/doe_pev_consumer_guide.pdf**)**

Searching for Information

We previously discussed studying the feasibility of an invention naturally requiring information that help evaluate it. Information such as users, the need for the invention, regulations and cultural limitations, etc. You cannot have a proper evaluation based on speculation and guesswork, even if guessing can be part of the process but it is at the very bottom.

Information sources are divided into two: main sources and secondary sources. Main sources consist of information from its original source, while secondary sources are a retelling of the original source. For example, a specialized center released a report about a detailed study of a given topic. This report is a main source. Then, when books, magazines or websites spread a part or all the report, this would be a secondary source.

Therefore, receiving information from main sources is the better option because it is more precise, as secondary sources may be missing something or have mistakes in them.

There are ways to gather the necessary information for an invention feasibility study using the criteria provided previously. These ways are:

1 – Searching the internet for a similar invention. Search the name of the invention or the name of the problem. This requires having searching expertise, as the internet is an endless world of information, and it will be quite difficult to acquire the precise information you seek. For more information regarding this, please check appendix (2) at the end of this book.

2 – Searching websites of patents or visiting patent offices in your country.

3 – Ask around. Ask people that are likely to need the invention. Ask them about how much they need it, and the price they are willing to pay for it. You could also conduct paper surveys, distribute them to study the need of the invention, while being careful of having your idea stolen.

4 – Inquire at shops about the availability of a tool or device similar to the one you would like to invent. Do not attempt to explain your invention. Only try to ask about a device that does the task of your invention. This is in order to avoid having your idea stolen as well. If the shop's answer is no, then ask them about where you could find such a thing. This could help you find more leads to work with.

Gathering invention information is a long and extensive process, and it takes a lot of time and effort. Modern means such as the internet, e-mail and others have made it much easier for researchers to perform this process. Twenty years ago, these means were not available, and researchers would put in a lot more effort to obtain information for any project.

Chapter 4
Technical Aspects of Inventions

After concluding the feasibility study for an invention, and measuring the expected success of an invention, comes the stage of turning the concept to a reality. This is often a part of the study process because of the need to create prototypes to run tests with it. This way, we get to know the downsides of the invention and proceed to accordingly enhance it.
Technical experience required differs from one invention to another, from one field to another. For example, the experience required to create a simple electric device differs from the experience required to create a complicated device that has applications on it. Inventing different types of medicine requires experience in chemistry, and so on.
On the other hand, there are things that are shared between inventions, such as their ability to be converted into a product sold commercially.
In this chapter, we will discuss six aspects of this topic:

- Technical Experience
- Gaining Experience
- Documentation
- Risks
- Turning a Concept into Reality
- Product Requirements
- Product Description

Technical Experience

What kind of technical experience to create an invention? Carpentry, painting, ironworks, electricity experience, etc.?

No invention can be made without the availability of technical experience. It is required not only at the stage of turning a concept into reality, but even before that, at the stage of coming up with conceptual designs on paper. For example, if an invention has electronic circuits in it, the experience to handle something of that field must be present. This in order to be able to draw the parts and choose the suitable materials and calculating any necessary values. Likewise, if an invention consisted of mechanical parts, there would be gears and reels involved. These would need careful calculations as well. If the invention was a medicine or a chemical component, it will be necessary to know the characteristics of all the different components within, and so on.

Do not hesitate to consult experts or to research books or the internet in the event of a lack of experience to create the invention.

The technical aspect of an invention goes through several stages as seen in figure (4-1). It begins, as previously mentioned, with conceptual designs on paper, and from here, you will need technical sources that inform you about the characteristics of each material you may use. These sources are made by the manufacturing companies of those materials and they are likely available on the internet. For example, a company such as RS is a distinguished company in selling thousands of different electronic and mechanical parts, as well as different devices and equipment. You can check their website with the following link https://uk.rs-online.com/web/. Another example is Premier Farnell, another company that sells thousands of parts and devices. You can find them in https://www.farnell.com/. These are just a couple of examples, there are many other companies that do the same and more.

As for the second stage, it is turning what is on paper into something real. This process is a verification of what was there in the first stage. Trying to work on the conceptual design could need a workshop or a laboratory with the necessary equipment that will aid you in its creation. This will need materials, and some materials may not be available in local markets, so you would have to search for them online and import them from outside, or find an alternative for them. Try to make use of any old device that is no longer functioning, they may have parts that could be turned into something useful for your current project.

An invention could have several aspects, and you may not be familiar with all of them, but that is no cause for concern. In a situation like this, you can consult an expert or ask for the assistance of a workshop to help you with what you lack. For example, if you need to create a wooden box, or a metal or plastic part, you could do that at a workshop.

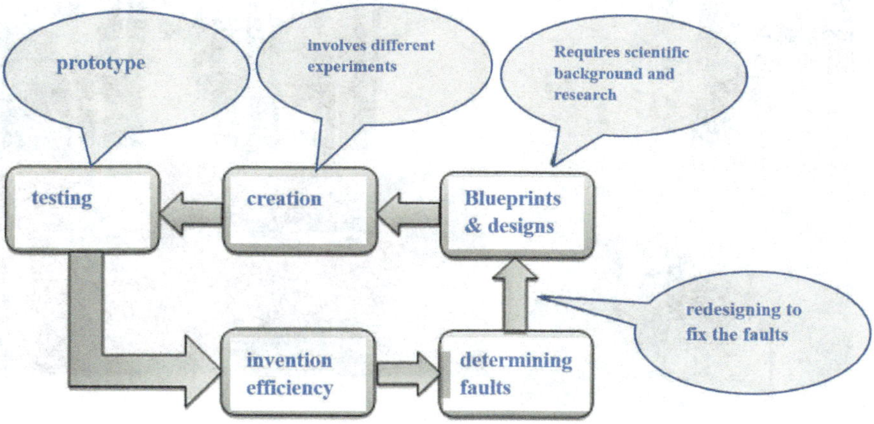

Figure (4-1) Stages of invention creation

Through the creation process, you may discover numerous mistakes in the conceptual design, such as incompatibility with some of its parts, or that some parts may have deficits, but trial and error will solve all these problems. You may find out that a conceptual design is essentially incorrect which will require going back to step one and making a new design.

The third stage is after the creation is finished. You can then proceed to test the product and its efficiency, faults, and make sure there are not any side effects for it. When you test your invention, you are verifying its eligibility to be pushed out into the markets.

Testing an invention's efficiency can be done in many ways, but it depends on what the invention is and where it is used. For example, you could try to measure its endurance in different temperatures. You could try to measure how much pressure or vibration it could withstand. You could try other tests with water leaks or its effects with chemical components or magnetic fields.

In the next picture, we will see special equipment used to test different kinds of furniture. These equipment are used by furniture manufacturing companies to measure the weight these seats could withstand.

Figure (4-2) shows a machine applying pressure to a seat, in order to determine how much weight it can endure.

Footnote: Some new products failed to sell in the market because they did not run enough efficiency tests.

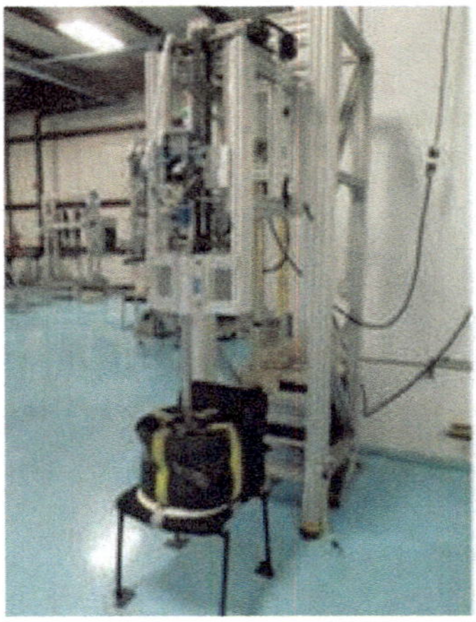 

Figure (4-2) Two types of machines used to measure seat endurance (Source: www.furnituretest.com)

As for figure (4-3), it shows another machine, but this one is used to test office chairs. By using a belt tied to the back of the chair, the machine can measure its strength by pulling on it multiple times, in order to test the springs that returns the back rest to its original position after a person leaves the chair.

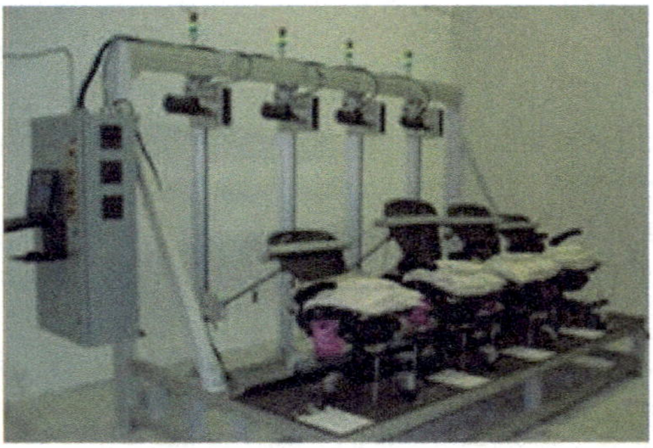

Figure (4-3) A machine used to test the endurance of an office chair's back rest (Source: www.furnituretest.com)

In figure (4-4), you can see the same aforementioned device, but this one testing the armrests

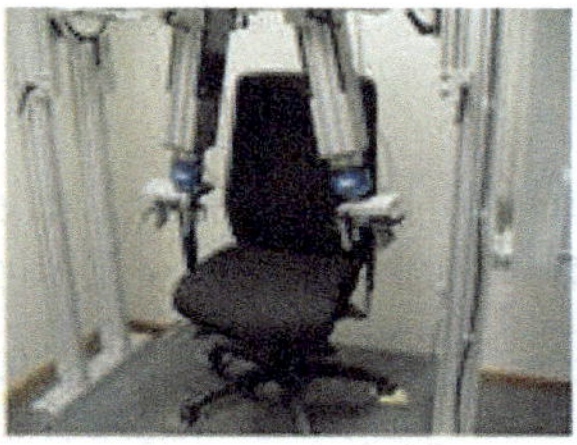

Figure (4-4) Armrest endurance test machine
(Source: www.furnituretest.com)

As for the machine used in figure (4-5), this one tests closet door hinges. It would repeatedly open and close the door to determine the hinge lifespan.

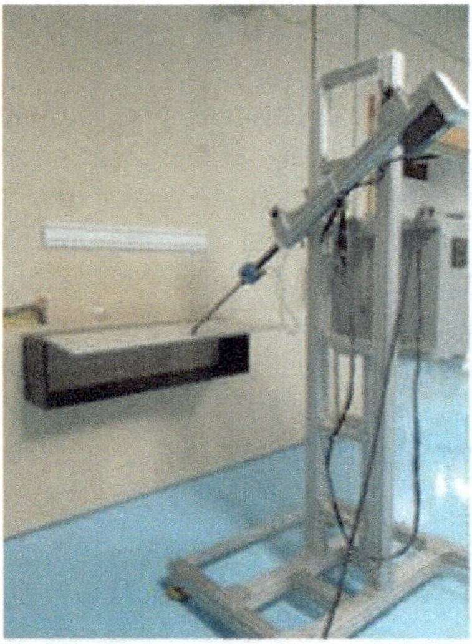

Figure (4-5) Closet door hinge test machine
(Source: http://www.furnituretest.com)

Based on the results of efficiency tests and determining faults, we proceed to correct the design and re-test the invention, all to make sure the removal of any faults. This will continue until you cover all the faults, and even after doing that, consumers may discover new faults. The product going out into the market and being used by people will reveal to us what the tests were unable to reveal. Therefore, some companies resort to testing products by giving them to volunteer consumers before shipping out a product and selling it commercially.
The trial-and-error process may take an extensive amount of time and an arduous amount of effort, depending on the type of creation, and the availability of information and materials and many other factors.

Gaining Experience

How can we acquire the necessary experience to grasp the solution?
Grasping the solution, in other words, the invention, requires many tests and experiments. This is in order to prove scientific facts that the invention will be based on. This is what we call an experimental course.
Scholars reached the consensus that Muslims' innate abilities with science stem from them embracing the policy of "Experiment and observe before having an opinion". Jaber Ibn Hayyan had always told his students not to neglect scientific experiments and to only rely on them, with careful observation. Another piece of his advice says "The first order of business is to conduct experiments! Those who ignore this step will not reach even the bottom levels of mastery. You must experiment to gain knowledge." (Source: An article of Ragheb Sergani from January 14th 2009)

Figure (4-6) shows that gaining experience and knowledge is what helps reach a solution. It is a series of new information never seen before. In order to gain experience, different experiments based on past experiences must be conducted. Making use of others' studies and past research is also necessary, as scientific research begins from where others have stopped. A lot of inventions began their first experiments by a certain individual and then later, another individual would pick up where the previous individual left off and build upon it further. This has been happening for many years now. Each individual would gain something out of past experiences and it would help them find their solution.

Figure (4-6) stages of reaching a solution

Conducting experiments also involves testing many materials. This is in order to know what is compatible with the invention. We must also acknowledge that a material used in the prototype may not be suitable for the final product. It may be too weak, a component in it could change with extended use, it could be unsafe, etc. Your search for suitable materials may lead you to invent something new or at the very least discover new things.

Conducting experiments also entails testing different methods in order to find the most suitable approach. For example, there may be a component that is incompatible with another, or may have side effects, or there could be a need to change the quantity of a material or it may be too powerful. All of this is part of the trial-and-error process. It is at times tedious and tiresome; therefore, patience is necessary. A valuable advice we could learn from history is the advice of Jaber Ibn Hayyan:

1 – Be patient, diligent, cautious and silent.
2 – Avoid the impossible and the pointless.
3 – Do not judge a book by its cover; this will likely lead your experiment to fail.
4 – Scholars should not be proud of the quantity, but with the quality of their work.

"You must act slowly and deliberately, and follow the trail of nature in everything you seek."

Source: An article of Ragheb Sergani from January 14th 2009, in http://www.islamstory.com/ar)

In order to elaborate, we will talk about an example. Let us assume that the invention required is a medication for a certain illness. After analyzing the illness and recognizing its causes and its effects on the different parts of the body, a conceptual design and calculations were made, and it was decided that four different components would be used, component A, component B, component C and component D. We must ask the following:

- Are these materials truly the ones needed to create a medicine that is potent enough for the illness?
- What is the required percentage of each component?
- Will there be side effects on the patient?

In order to get answers for these questions, we must conduct experiments and tests to obtain the evidence and information that prove them.

Another example which is real: A story of the Wright brothers with their invention of planes and how they obtained the necessary experience to solve their problems.

Not every experiment conducted by them was successful. Out of seven airplane prototypes that they built, only two saw success. This was between 1899 and 1905 as shown in figure (4-8).
The Wright brothers together conducted experiments on 200 different designs for the wings, by using propellers as shown in figure (4-7), this was in order to know which design was the most able to lift a plane.

Figure (4-8) An airplane prototype

Figure (4-7) Plane propeller

(Source: http://www.wright-brothers.org)

The Wright brothers did not only search for the most suitable wing design, but they also designed the propellers by themselves. A propeller is a long tube with a fan at its end, when the fan spins, it creates airflow that goes out from the other side. The brothers used propellers as a means of studying airflow on the wing and the plane body, and how to control it.
The Wright brothers tried different prototypes in order to know more about the airflow effect on the plane's body, and how to control it through things like flippers as shown in figure (4-9).

Figure (4-9) Wright brothers' experiments
(Source: http://www.wright-brothers.org)

The brothers studied all the experiments past scholars have conducted. The most prominent one was by Andalusian scholar Abbas ibn Firnas. He was the first to experiment with aviation, and that was by making two wings for himself. The brothers also studied the behavior of birds and concluded the importance of the tail in landing and balancing. This is precisely what Abbas ibn Firnas had pointed to as well in the 19th century. (Source: A book titled "Muslim Heritage in Our World") The fall and crash of some plane prototypes as in figure (4-10) was a source of many informative scientific facts. Without them, human beings would not know so much about aerodynamics. There is, therefore, a benefit to experimenting, and that is obtaining new information. Through explaining the reasons of failure, and conducting a new experiment, we can eventually reach a solution.

Figure (4-10) The crash of a plane prototype
(Source: www.centennialofflight.gov/essay/Wright_bros/1900_Gliding/WR2.htm**)**

Documentation

Documentation is an essential part of scientific research and discovery. Every process of an invention must be documented in as much detail as humanly possible. You must have your own invention notebook where you write everything you have worked on, from the inception of an invention idea all the way through experiments and then the end of the project. It will assist you in many things, such as revision, making sure of some steps and their results as well as proving all your efforts. This will help you when you apply for a patent. If it was possible to record footage of your research, then that may help you create a documentary about the invention.

- Documentation may involve points we have previously mentioned in Chapter 2, which we will mention once again:
- The name and description of the invention. This is the first thing you should document as it is the first stage. It can also include how the idea was made.
- Concept art for the invention: Concept arts are extremely important.

Projects that involve engineering or science must have drawings. Text alone is not enough. Draw everything as clearly as you can, and you can also make use of some computer applications to help you draw.

- Required parts and materials. You must write down the names and technical information of every parts and materials you will use, as well as their sources, prices, etc.
- References. An inventor must not take for granted any technical references that are relevant to their invention. This includes books, websites on the internet or even people of experience. So, make sure to document the details of all these as well.

Risks

Working on inventions and scientific discoveries, as important as they are to human life, is not without its dangers. You must always remember to be cautious while working to avoid anything you would not want to happen. For example, if your invention has flammable materials or high voltage electrical currents and the like, make sure to have safety measures.

Many scientists and researchers throughout history paid the price of safety negligence with their lives. We will mention some examples here with the purpose of benefitting from the experiences of others.

1 – Karl Wilhelm is a German chemist and pharmacist. He was born in 1742. Many discoveries of chemical components were thanks to him. He usually would taste the components that he discovered. This led to his demise in 1786 after tasting mercury and poisoning himself with it.

Figure (4-11) Karl Wilhelm

Source: (www.incrediblediary.com/10-scientist-who-killed-themselves-by-their-experiments/**)**

2 – Elizabeth Ascheim (1859 – 1905) was working as a librarian. She left her job after taking an interest in studying X-rays, which were discovered by Karl Wilhelm, with her husband who was a doctor. She brought a machine for X-rays to her husband's office, and this would be the first laboratory for X-rays in San Francisco. In 1897, the couple began experiments on themselves without considering the risks. Extended exposure to these rays without protection would inflict cancer, and this led to Elizabeth's death.

Figure (4-12) Elizabeth Ascheim

Source: (www.incrediblediary.com/10-scientist-who-killed-themselves-by-their-experiments/**)**

3 – Sir Humphry Dav (1778 – 1829) was a British chemical scientist known for his intense love for chemistry. He used to conduct many experiments and grew to enjoy inhaling different kinds of gases that he was working on. Even though this habit assisted him in discovering characteristics of the sedative nitrous oxide, it was the same habit that caused the downfall of his health after inhaling poisonous gases. Sir Humphry also got blinded because of an explosion of nitrogen trichloride during one of his experiments.

Figure (4-13) Sir Humphry Dav

Source: (www.incrediblediary.com/10-scientist-who-killed-themselves-by-their-experiments/**)**

There are many other stories regarding this topic. From what we have just read, it is now clear that dealing with new and unfamiliar territory may have undesirable consequences. We must therefore, as inventors, always be cautious. To avoid falling into similar unfortunate situations, here are some tips that could help you stay safe during experiments.

1 – Before conducting any experiment, try to obtain as much information about the materials as humanly possible, particularly with chemical components. You must also be knowledgeable about how to use the different tools you have.

2 – Consult different experts and heed their advice before conducting any experiment, or before even trying to use any equipment to learn or to take out parts from it. Some machines may have sharp parts in them that could cut you. Some electric devices may have electricity in some parts after turning off the power, like with ovens and microwaves. They contain high voltage electric capacitors as shown in figure (4-14), and other unexpected sources of danger.

Figure (4-14) high voltage converter on the right, and a high voltage capacitor found in microwaves on the left.

3 – The workshop or lab that you work in must meet all safety requirements. This includes having a safe experiment table that ventilation for harmful smoke or gas as a result of chemical reactions. This would lead them outside the lab as shown in figure (4-15).

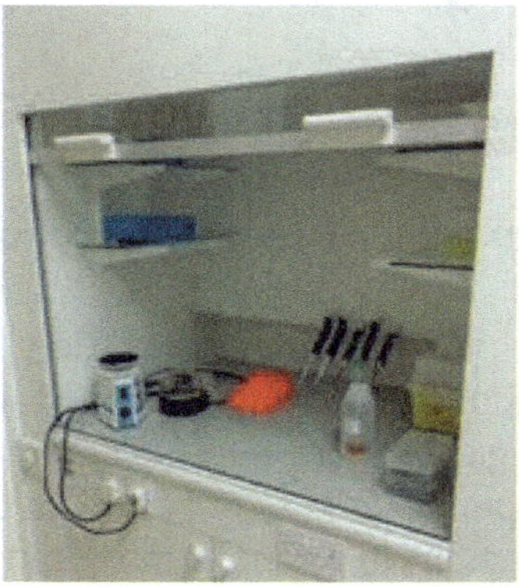

Figure (4-15) Safe experiment table.

Having fire extinguishers and tools related to it as shown in figure (4-16) for any potential fires is also extremely important.

Figure (4-16) Fire extinguishing equipment

se safety goggles, masks and gloves like the ones shown in figure (4-17) as well if necessary. Storage for chemical components must also be well-stockpiled. Make sure that movement will not be difficult in the lab. On top of that, having a first aid kit around is of utmost importance.

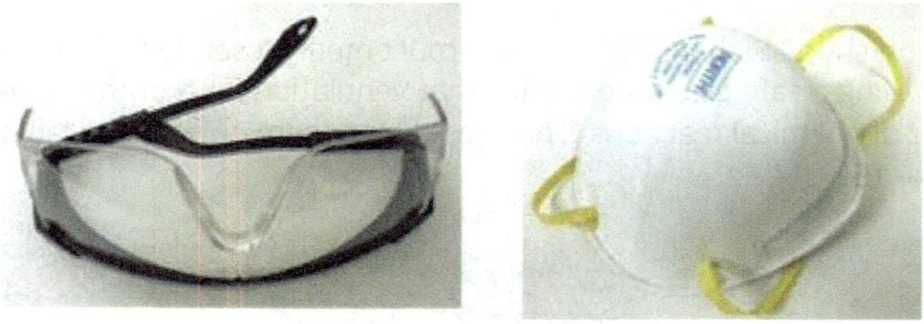

Figure (4-17) Safety equipment such as a mask and safety goggles

4 – In case of an invention containing flammable gases or liquids, make sure to use explosion proof equipment as shown in figure (4-18). They are made in way that prevents the leak of gas to them, especially with Pressure switch and Relays, contacts could cause a spark that may not even be visible to the eyes, as shown in figure (4-19), but having gas leaked into them can cause an explosion or at the very least a fire.

Figure (4-19) A spark caused from Relay contacts

Figure (4-18) Explosion proof pressure switch (Source: www.microlectra.com)

Turning a Concept into Reality

Making an invention tangible starts by putting a conceptual design. Drawings that demonstrate the parts of the invention, and the sequence of the stages of the invention until the last. The last stage is the invention's delivery of the desired tasks. Each stage is a certain process, and planning all of it requires engineering that involves calculations such as height, width, capacity, weight, size, ability, the magnitude of the electric current, etc. It also involves determining the required tools and materials.

There may be several designs and ways to carry everything out, so how do we choose the most suitable approach? It all comes down to the lowest-cost and easiest option.

Design Stages

You must recognize each stage that your invention goes through until it reaches the final stage as shown in figure (4-20). Keep in mind that the number of stages is indefinite because it depends on the nature of the invention and the kind of task that it performs.

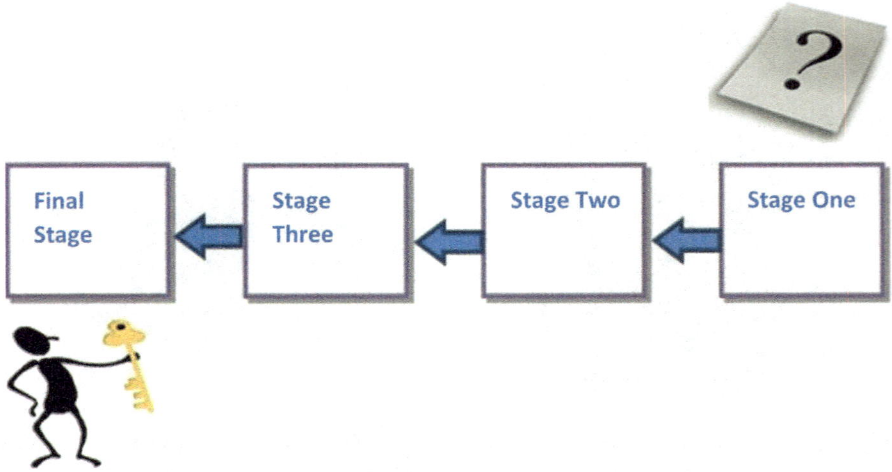

Figure (4-20) Sequence of invention stages

To make things easier, we must answer the following questions:
What are the required tasks from the new machine or device?
And what is this invention supposed to give us? What is the function of it? This should be determined since the inception of the concept.
What are the stages of the invention's creation? How do we organize those stages? This requires thorough planning, just as you would while driving your car to a destination. You will more than likely try and find the best route that takes the least time and is the least crowded.
How many fields does the invention combine? Mechanic, electric, electronic, etc.
Knowledge and technical experience are essential and you should not hesitate to consult experts in fields that you are not well-versed in.

To explain everything, we will mention a simple example with an invention that we would like to make. Let us take for example a tea-making device.
How can this device be made?
It is obvious that the result of the final stage will be receiving a cup of tea, but what about the other stages?
To make things easier for you, put yourself in the place of the device, what would you do if you were asked to make a cup of tea for someone?
You will simply take a cup, add a spoonful of sugar or more, depending on the requested amount, then you add a tea bag and lastly boiled water. That is all. This is precisely what the machine will do, and we can summarize it in the following table:

Stage	Process	Quantity	Notes
1	Getting a cup		
2	Putting sugar	1 spoonful or more	
3	Putting tea bag	1 bag	
4	Pouring boiling water	95% of the cup size	Water heater required
5	Delivering tea cup		

According to the order of stages, we can see that getting a cup is logical and it must be the first step. As for step 2, 3 and 4, the order does not matter as much, it could even be all at the same time. After determining the correct sequence, we begin adding a conceptual design for the creation of this machine.

We will need three containers, one for sugar, one for the tea, one for the water. We also need electric valves to control the flow of sugar, tea and water into the cup. We should consider that the amount of sugar will increase or decrease at the request of the consumer. The water container is a kettle that has a controllable temperature as shown in figure (4-21).

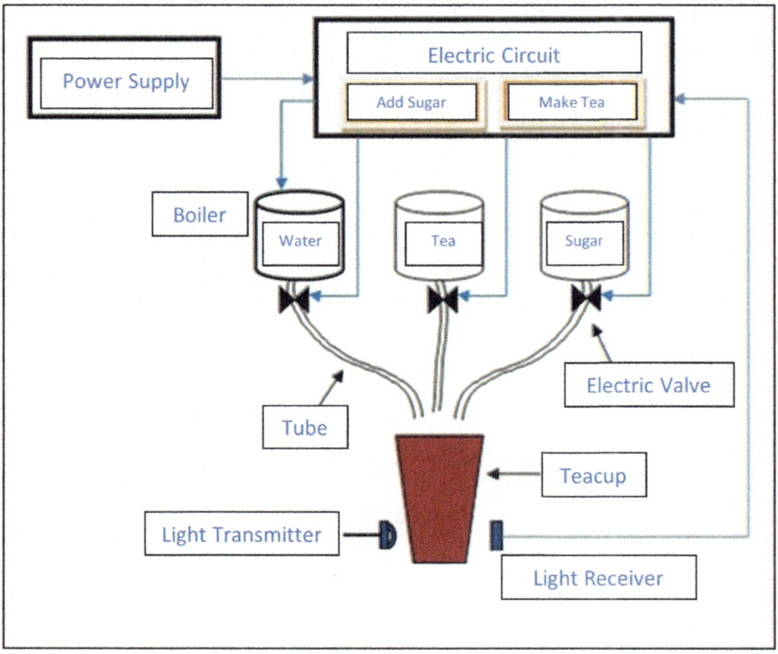

Figure (4-21) Conceptual design for a tea-making machine

After pressing the "Make tea" button, the electric circuit opens the electric valves for a certain period to allow the passage of sugar, tea and boiled water into the teacup. The time for every valve differs from one to another depending on the required amount of sugar, tea and water. This time would be previously programmed into the electric circuit. As for the amount of sugar added, that can be solved by adding an "Add sugar" button. Once pressed, the sugar valve opens up for a longer period of time to put more sugar into the cup.

The electric circuit is also provided with a thermostat to control the temperature of the kettle. As for the cup, it is placed by hand under the three tubes. Naturally, a sensor is required for the circuit to determine the placement of the cup. The light transmitter and receiver can be used to help with that purpose. As seen in figure (4-21), when putting the cup, this gets in between the light transmitter and the light receiver, which sends signals to the valves to open to make the tea. If we wanted a machine that puts the cup by itself, then we would have to make a mechanical device that controls the circuit, but this, of course, is more complex.

Lastly, there must be a power supply to power the electric circuit, the valves, and any other components.

After drawing the conceptual design, we move onto searching for the parts such as the valves, the kettle and others. We must also determine their technical specifications, the electric circuit and the power supply. Will you make them by yourself, or will you look for something pre-made?

Product Design Requirements

Obtaining a patent does not necessarily mean that your invention is usable and ready to be manufactured and sold. There are several aspects to take into consideration in order to make sure the invention will be viable. This is what we refer to as "product design".
Designing different products is a scientific discipline. It is a field that has its experts that recognize the different aspects that a product must contain. Designing products is a field that links a multitude of things together, such as art, science, safety, etc. All in order to make certain that the invention is a usable product. Designing a product does not only entail making sure of its functionality, but it also has other aspects such as safety, ease of maintenance, how to package and ship it out, etc. The designer must be aware of what suits the consumer and their circumstances. We will discuss some of those aspects.
In order to attain the best design for your product, it may require trying out several different designs and choosing the most suitable. It can take a lot of time and effort, as every design you come up with may have faults that may result in the product's failure, but never despair, keep trying, and always try and benefit from each experiment you conduct.
We have all used, or at least know of, the renowned WD-40 spray, shown in figure (4-22). This is used to remove rust from screws. This chemical component was made by Norm Larsen after 39 failed attempts, in which he succeeded at the 40th. This was the secret to the addition of the number 40 to the product.

Figure (4-22) The renowned WD-40 spray

1 – Maintenance Aspect

The product designer needs to have a deep foresight for their product. It should not end simply after the product leaves the manufacturing factory and drops into the markets. The product may need some maintenance work after time, and maintaining it will be difficult if the designer did not consider this aspect of the product while designing it.

We all notice this in our daily lives at home or at work. For example, when we try to replace a damaged light, we find difficulty in removing the light stand or to put it back on, look at figure (4-23). To replace a broken light stand with another, you must remove the glass cover, because the hole is tight and it is difficult to get your hand in. To remove the glass cover, you need to remove the base from the wall. After replacing the light, you will put everything back in its place once again, and it is, of course, a very inconvenient process.

Figure (4-23) Difficulty in replacing a damaged light

There are consumers that consider the maintenance aspect while purchasing their tools and devices, and there are others who do not pay attention to this aspect and end up discovering these faults and will refuse to buy the same product again, and they will spread the word to their friends and relatives.

Another example of this aspect is the use of cables on a wiring board of different sizes. They should be placed carefully in a specific way in order to know where they belong in case of having to remove any of them during maintenance as in figure (4-24). There are some cables that could be used both ways, and these are not recommended because there are many instances where the cable is unintentionally reversed. This could lead to damage in a device. Use mono cables as shown in figure (4-25).

Figure (4-25) Mono cable

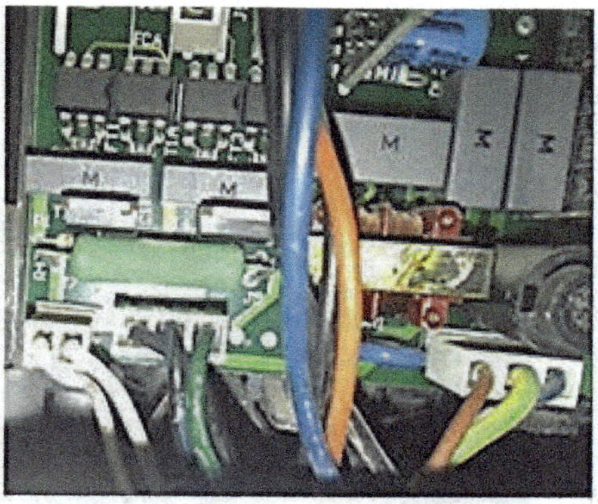

Figure (4-24) Types of cables on a wiring board

If you were ever a technician, and you previously opened some electronic devices, you would have noticed this well.
Figure (4-24) shows three cables. Two of them are similar in appearance, and the number of the cable is written on the wiring board. This number is not written on the cable itself. If you remove the cables without placing any kind of marker later, then you will find yourself troubled.
Removing a cable and putting it back together in the wrong place, or at the right place but reversed during the process of maintenance could lead to device damage.

2 – Upgrade Aspect

Science and technology are ever-evolving, especially when it comes to devices that contain applications and programs such as personal computers, control systems, monitoring, etc. Upgrading two aspects, hardware which are electronic pieces like memory and processing units. The other aspect is software, which operates the device's functions.
As for hardware, it works by implementing a default setup on the wiring board as shown in figure (4-26), where we, at any point of time in the future, can add any new part to the device or replace it with another that is better, as shown in figure (4-27), and it can be done very conveniently.

Figure (4-27) Some parts can be attached or replaced easily

Figure (4-26) default setup to add another part in the future

As for programs, this requires the device to be connected to the internet in order to download the latest update. If a device needs a software upgrade and is not connected to the internet, then you could download the update on a laptop and then proceed to transfer it onto the device via a cable. Some devices are unable to access the internet, such as scales, control systems or some medical devices, and laptops can help with that as well. The invention designer should keep in mind the implementation of a computer port so that you can easily connect a device to a computer without the need to open the device as shown in figure (4-28).

Figure (4-28) a RS232 port that connects to a computer on the back of a satellite channel receiver.

3 – Safety Aspect

Safety is one of the most important aspects that should not be neglected when designing a product. The usage of any product must not result in any sort of harm to the consumer. Parts with high electrical power should be covered well to avoid any electric shocks to the consumer or the technician performing maintenance. If there are thermal parts, then they should also be covered to avoid unwanted accidents.

A product should also not have sharp edges that could cause cuts on the consumer. This is especially evident in kids' games and moving devices. One time, I was working on the maintenance of a medical device and I cut my hand because of a sharp edge inside the device. The designer that overlooked this device's creation had overlooked the sharp edge under the device's cover is not safe for a technician or an engineer performing maintenance work on it.

Look at figure (4-29). It is a control panel that controls the doors of a building. The picture shows many examples of safety negligence. One of the batteries was put on the edge of another battery, and it could fall at any given moment. Falling would cause the wires connected to be cut and possibly cause damage to the panel. If the battery fell on the foot of a technician, they could cause a foot injury because of the weight of the battery. In addition to that, the figure shows the tangled wires and parts of the panel that were disorganized without stabilization. This is a prime example of bad design, safety wise.

Figure (4-29) the product designer did not take into account the safety aspects

4 – Outer Appearance

The outer appearance of a product can encourage the consumer to purchase it. The more stylish the appearance of a product, the more likely a consumer would be tempted to buy it. We all love and admire beautiful things. Therefore, we see companies manufacturing products specifically designed for children or for female consumers. There are types of cars and laptops that were decorated with colors that are more enticing for women.

As an example of the opposite to that, in 1957, the Ford Motor Company manufactured a new model of cars called Ford Edsel, shown in figure (4-30). The sales of this product were much lower than expected. One of the reasons is the unappealing appearance of it, as the lights were a bit high as if the car had eyes. In addition to the vertical radiator cover that many customers did not like in the US. (Source: https://en.wikipedia.org/wiki/Edsel)

Figure (4-30) Ford Edsel of 1957 that did not perform well

5 – Considering Climatic Conditions

In the 50s and 60s of the past century, European and American cars were the most common in the Arab Gulf states. Japanese cars, however, started to conquer the market in the 70s while European and American cars had their sales reclining. One of the main reasons how much Japanese cars suited the Arabic climate of high temperatures and humidity during the summer, whereas European and American cars were designed to suit the cold climate of Europe. European and American cars would struggle in the Gulf because of the high temperature because the radiator would not be able to cool down the engine. Moreover, the air conditioning was not enough for the weather of the Gulf. In fact, those cars would have seat warmers which are unnecessary in the Gulf.

The lack of or negligence even one of these aspects could lead to a failure of a product and the inability to prosper in a market.

Product Specifications

If you open the box of any new product, you will often find a paper attachment containing the specifications of the product. These specifications may be written in the instruction manual or even on the box itself.

Determining the specifications of a product is an essential matter. A consumer will choose a product based on what is compatible with the environment that they live in, or based on the need. Moreover, they will pick the specifications that are compatible with the specifications and measurements of their country.

The aspects of specifications of a product are different, depending on what the product is. An electrical product must have the voltage written, such as 220 volt or 110 volts, as well as the electric capacity that the product uses, such as 100 watts, in addition to the device weight, size, etc.

Wireless devices, in addition to the aforementioned aspects, must also contain the frequency type, and the maximum distance it could connect to a wireless signal. This is all important, as every country has a specialized committee that issues the permits that allow the use of communication devices to different sources according to the specific specifications that must be adhered to in order to avoid wireless frequency interference. The same goes for medicine and chemical components that have approved specifications in all countries.

Figure (4-31) shows an example with the specifications of a Samsung washing machine. In addition to electric specifications, there are specifications that are related to the amount of water, rotation speed, etc.

SPECIFICATIONS					
WA10U7					
Energy Consumption	Washing	470 watts		High	77 litres
	Drying	280 watts		Average	64 litres
Dimension		W601 x D654 x H968		Low	55 litres
Weight		36 KG		Minimum	43 litres
Water Pressure		0,05-0,78 MP (0,5-8 KG power/cm2)	Water Usage		212 litres
Washing Type		Movement Type	Motor Speed		720 Rpm
WA13U7, WA12U7					
Energy Consumption	Washing	470 watts		High	83 litres
	Drying	280 watts		Average	76 litres
Dimension		W601 x D654 x H993		Low	55 litres
Weight		39 KG		Minimum	43 litres
Water Pressure		0,05-0,78 MP (0,5-8 KG power/cm2)	Water Usage		227 litres
Washing Type		Movement Type	Motor Speed		720 Rpm

Figure (4-31) The specifications of 3 Samsung washing machines

As for figure (4-32), it shows the specifications of a Canon printer, such information on the printing paper, the colors, the speed of printing, etc.

9 Specifications

▶ General specifications

Power source	220V–240V AC, 50/60 Hz
Power consumption	• Maximum: Approx. 47.0 W • Standby: Approx. 14.0 W (MP360) Approx. 14.5 W (MP370)
Weight (Components attached)	Approx. 18.5 lb./8.4 kg
Dimensions	• 17 7/8 in. (W) x 14 1/8 in. (D) x 9 7/8 in. (H) (454 mm (W) x 358 mm (D) x 249 mm (H)) *With the multi-purpose tray and the paper output tray closed • 17 7/8 in. (W) x 21 1/2 in. (D) x 11 1/4 in. (H) (454 mm (W) x 547 mm (D) x 286 mm (H)) * With the multi-purpose tray, the paper output tray, and both tray extensions opened
Environmental conditions	• Temperature: 15°C–27.5°C (59°F–81.5°F) • Humidity: 20 %–80 %
Liquid Crystal Display (LCD)	20 x 2
Display languages	English/French/Spanish/German/Italian/Dutch/Finnish/Portuguese/Norwegian/Swedish/Danish/Slovene/Czech/Hungarian/Russian
Applicable standards	• Electrical safety: EN60950, GS • Emission: EN55022 Class B • Immunity: EN55024 • PTT: TBR21 • Other: Energy Star compliant, CE mark
Multi-purpose tray capacity	• Plain paper: approx. 100 sheets (20 lb. or 75 g/m^2) or 3/8 in. (10 mm) thick paper stack * For other paper types, see *User's Guide*.

Printable areas

• **Paper printable areas:**

A4 size: 8.0 in. x 11 3/8 in. (203. 2 mm x 289 mm)
B5 size: 6 7/8 in. x 9 7/8 in. (175.2 mm x 249 mm)
A5 size: 5 5/8 in. x 8.0 in. (141.2 mm x 202 mm)
Letter size: 8.0 in. x 10 3/4 in. (203.2 mm x 271.4 mm)
4" x 6" size: 3 3/4 in. x 5 3/4 in. (94.8 mm x 144.4 mm)
5" x 7" size: 4 3/4 in. x 6 3/4 in. (120.2 mm x 169.8 mm)

Borderless printing enables you to make prints with no margins. However, it may result in a decrease in print quality of the top and bottom paper edges.

• **Photo Paper Pro PC-101S printable area:**

Photo Paper Pro PC-101S: 4 3/8 in. x 6 3/8 in. (109.6 mm x 160.4 mm)

The printing area of Photo Paper Pro PR-101 is identical to that of A4-size paper.

▶ Copy specifications

Copy speed	• Black & white:	<FAST> mode approx. 18 pages/minute (A4 size)
	• Colour:	<FAST> mode approx. 12 pages/minute (A4 size) (Based on BJ printer specific speed test pattern.)
No. of copies	• Maximum 99 copies	
Density adjustment	• 9 levels	
Zoom	• 25% to 400%	

▶ Photo print specifications

Interface	• Card slots • USB port
Print media	• Plain paper • Photo Paper Pro PR-101
Layout	• Plain paper: A4/Letter: With/without borders, index (max. 70 images) • Photo Paper Pro: A4/Letter: With/without borders, index (max. 70 images) Card: With/without borders, index (max. 15 images)
Features	• Date printing • Vivid photo printing • DPOF function-compatible (Index printing, specified quantity, specified images, date printing, image number printing) • Digital camera direct printing • Photo Index Sheet printing • Borderless printing

▶ System requirements

See *User's Guide*.

Figure (4-32) Canon printer specifications

Chapter 5
Manufacturing and Marketing

How can you turn your invention into profit? That is the end goal that needs a well-planned and thorough plan as well as doing all that is within your capacity in order to achieve.
If you do not take the initiative to convert your invention into a profitable commercial product, your concept that you have worked so hard on and cost you a lot of time and effort would all be for nothing. It is important as an inventor to know the available options of marketing your invention. Any profits you get can be used to invest into another invention.

Without the existence of an effective plan to manufacture and market the product, you can never profit from your invention, even if the invention is a big or an important one. Achieving this requires the organization of a team that consists of three pillars. Besides you, as the technician, inventor and designer of the new product, there will be a need for a manufacturing expert that understands the different aspects of creating a product as well as choosing the right materials bearing in mind the cost of manufacturing. You will also need a marketing expert that knows the most efficient promotional channels for the product and how to reach consumers and determine the market competitors. You can later add other pillars such as a product designer separate from the inventor and the engineer, and other pillars depending on how much the product will grow.
The members of this team must work together, and you, the creator will be the link that connects everyone together, because the end goal is to produce a suitable product of decent quality, sold at a reasonable price, and therefore gaining profit.
In this chapter, we will discuss multiple options available for manufacturing a product, and we will also discuss some ideas about marketing.

I: Manufacturing

Manufacturing any new product may not be an easy task, and may entail numerous challenges. Any new product is prone to failure because of lack of experience in the different processes of manufacturing such as choosing suitable materials, or a suitable design, or even the proper packaging for shipments, etc.
Quality assurance must not be neglected to reduce the production budget, and vice versa. There must be a balance struck between the two. A consumer would not want an overpriced product, and at the same time they would not want a cheap product with a mediocre quality. For example, you as a consumer would not want to buy a cheap electric light that burns every two weeks, and you would have to buy a new one. You will be ready to pay double to buy a light that lasts for a month or more.
In some fields, the quality is the first thing to be looked at, such as with medical machines. Medical equipment is very delicate, as the lives of patients in hospitals rely on them. It is simply not worth risking either the price or the quality in this instance. The same goes for computers and manufacturing control systems in factories and oil refineries. The halt of production in a factory because of a malfunction could cost them thousands.
Manufacturing products has changed since the beginning of the 20th century. In past times, there would be a reliance on the efforts of one person or a small group. But with the beginning of the 20th century, production lines were upgraded which allowed production and manufacturing products at larger quantities. A production line is a series of consecutive stages in a factory, where each stage adds a part of the product until it is fully made at the final stage. Some stages may be by hand done by real people and others may be done by machine, but it is usually machine-made amid all the advanced technology. By the middle of the 20th century, sales and promotions for different products were also developed.
As for the process of turning an invention into a commercial product, there are several choices available that allow you to make profit. You must first make sure that you have registered a patent for your invention, as it is essential to have because of the value it adds to your product in the market. Let us now look at the options that allow us to turn your invention into a product.

Footnote: On December 1st, 1913, Henry Ford established the first production line for car assembly. This was revolutionary for the cars industry and it re-defined manufacturing around the world.

Option one:

You can sell your invention to a company. They would handle the manufacturing and marketing, assuming you have the patent. This requires you to search for a company that is willing to buy it from you, and you will need to be persuasive to convince them about the importance if your invention. Companies will not invest in what is not profitable for them.

You must also be cautious so as not to have your invention stolen. Therefore, you must only offer your invention to trusted companies. You may have them sign a confidentiality agreement as a precaution, and make sure to consult legal experts regarding this matter.

Option two:

Bring the invention to a manufacturing company outside of your country, and then you handle marketing. You must also be careful here to not have your invention stolen. Make sure to always communicate through the proper, trustworthy channels. You may also use contracts to help you in the event you get into any sort of legal conflict.

One way to avoid leaking the secrets of your invention, is to manufacture different parts for your invention at two different factories and then you assemble the parts by yourself. This will prevent you from confiding all details of your invention to one factory.

Option three:

Give a manufacturing and marketing permit to other companies by signing a yearly contract in exchange for a percentage of profit while keeping the patent for yourself. This is considered a convenient option, because compared to the fourth option, which we will discuss shortly, you will not need to exert effort and spend time in the establishment of a factory and everything that it entails such as manual labor, follow-ups, expenses, calculating profits and losses, etc. In the first option, you will gain your profit as a one-time payment and that is by selling your patent, whereas with the third option, you will continuously gain an amount without any effort. Profit with this option may be less than the first option's immediate payment, but it will be constant.

Option four:

Manufacturing and marketing the invention by yourself, as in, establishing your own company. How can you do this and what will you need?
Manufacturing the invention by yourself simply means making your own company, but what exactly will you need to be able to do this?
Manufacturing a product has some positives on one hand and obstacles on the other hand. It will entail many administrative matters that you must learn and even master in order to manage this factory. Making the new product is not the only thing there is to do, there are other things that entail such as marketing and promoting the product. These aspects have their strategies that need experience and practice, you will also need to learn about managing your workers, monitoring them, etc. In short, you will have to establish a company, and the factory will only be a part of it.

This factory can be a place to simply gather different parts together for the product, and then you would take these parts to other factories and give them your specifications that you want to implement according to what your invention needs, and this is what happens with car factories. Many car factories are an assembly line for different car parts, and these parts are bought by other companies that are more specialized, and they would build them based on the desired specifications. For example, there are companies that specialize in manufacturing car lights, and others that are specialized in tires, and others specialized in specific parts of the engine, and so on.

Bearing that in mind, you may face obstacles such as not being able to find a company that wants to manufacture parts for you, or at least not in the way you want them to be. This is likely because many smaller factories like metal, plastic and fiberglass workshops always create the same things, and they are not used to creating new things that they are not used to, with rare specifications. Because of this, you may have to pay a higher price to get what you need. This will naturally raise production costs and affect competition in the market. Either this, or you would have to compensate with the quality or change the design. This may take away important features from your invention that may be important to compete in the market.

There is another important matter as well, and that is the extent of the workshop's ability to provide you with parts that you need for your invention.
This workshop may be one that you rely on, and guaranteeing that they continue to produce the parts needed is guaranteeing the continuity of the manufacturing of your product. These parts become raw material for you, and if they stop providing

them, then your production will certainly cease. In order to avoid this, have a contract with more than one parts provider, and have a sufficient supply of raw materials so that you guarantee that you will not run into a situation where the production ceases because the production of raw materials is late, or having delays in shipments, etc.

Evaluating Providers

The flow of raw materials to your factory without halt is essential. Therefore, you must evaluate the parties that provide you with these materials, whether they were companies, small workshops, etc. This must be done before the contract. Here are some important evaluation criteria to keep in mind:
Quality: The provider must be consistently precise with the quality of the parts they make for you. In other words, they must be able to make these parts with the exact specifications that you need without any faults. Therefore, you should not accept a batch with inconsistent material specifications.

Price: The provider that does the job for you at the lowest price is the best option. This will help reduce production costs and allow you to compete in the market. This does not, however, mean that we should choose the lowest price and ignore the quality required. As previously mentioned, your product relies on the quality and the parts you use.

Service: Are the providers able to provide you with the required materials at the designated time with the requested amount? Are they ready to give insurance on the materials in the event that there are some electronic or mechanical parts? Your product may stop functioning during the insurance period with a consumer, and you would have to compensate the consumer for that. If the provider can compensate you as well, that would be even better.

The three criteria points aforementioned apply to the second and third manufacturing options, where choosing the company that manufactures and markets your product must only happen after you evaluate them, in order to pick the most suitable company that pushes your product into competition and success in the market. The more reputable the company you pick is, with good customer service, good administration and promotion work, the better it is for you. It would be a waste of time to work with a company with unenthusiastic and uncooperative administration, and that can be a reason for failure.

SWOT Analysis

Whether a factory is a simple assembly plant or a large one, a feasibility study must be deliberately conducted without the need to rush the process. This is in order to measure the success and the potential of establishing an industrial organization. There is no end to the number of commercial projects that ended up in failure time and time again, and accumulated debt for their creators because of the lack of a feasibility study and analyzing the expected opportunities and obstacles. In this regard, we will discuss the well-known SWOT analysis. This is a tool to measure the potential success of any commercial project or even non-commercial ones.
The SWOT analysis is a famous theory in administrative studies, and is used to plan projects to know how successful they may be. This is done through analyzing four central points that decide the success or the failure of any company or establishment. It is a tool that helps you decide.

SWOT is an acronym that stands for: **S**trengths, **W**eaknesses, **O**pportunities, **T**hreats

These are the four aspects that the analysis focuses on. If we investigate the elements of each of the four aspects, we will be able to determine how successful a project can be.

A Brief History of the SWOT Analysis

The history of this analysis dates back to between the 1960s and the 1970s, when a team of researchers at the Stanford Research Institute in order to find out the reasons behind the failure of companies in their planning and strategies. A total of 500 companies supported this research team, as they managed to study the problem with 1100 companies and establishments. After careful analysis and studying the results, they have concluded that the aforementioned four central aspects can help any company or establishment measure the success of their plans.
Source: (http://www.businessballs.com/swotanalysisfreetemplate.htm)

Applying SWOT Analysis

Take two opposing aspects and compare them with each other. You will have the strengths and weaknesses on one hand and the opportunities and threats on the other hand. The following table shows an explanation to these aspects.

Aspects:	Strengths and Weaknesses	Opportunities and Threats
Explanation:	Related to internal environment of the establishment	Related to the external environment of the establishment
Examples:	Aspects related to the product, whether it is technical, pricing, performance, quality, etc. It also includes factors related to the employees, their technical and administrative skills, etc.	Aspects related to the market and its variables and competition. It also includes rules and regulations of the country and all that is outside of the company's control.

We will demonstrate in the next table the factors or criteria that fall under each aspect of SWOT. Keep in mind that the criteria mentioned are not fixed and are subject to change depending on the nature of the project.

In this breakdown, you can clearly recognize all of the factors of the four aspects of SWOT, and work on improving them before working on the project by turning a weakness into a strength, and turning a threat into an opportunity. For example, if your budget is low, then you must find a source of funding that allows your establishment to have a head start in the market.

Internal factors		External factors:	
Strengths	**Weaknesses**	**Opportunities**	**Threats**
Positive product Technical aspects, Suitable product price, Employee efficiency, Big budget, Marketing ability, Good location, Quality assurance certificates, Low maintenance needed.	Negative product technical aspects, Unsuitable product price, Employee inefficiency, Low budget, Marketing inability, Bad location, No quality assurance certificates, Constant maintenance needed.	Many consumers, Low competition, Practical product, Banks financing the project, Official entities encouraging production.	Few consumers, High competition, Restrictions with local regulations, Banks not financing the project, Restrictions in hiring capable employees, Unsuitable climatic conditions, Unavailability of materials and tools.

For more information, please look at the SWOT analysis in the appendix at the end of the book.

Selling Your Invention Online

There are websites that promote the idea of selling patents. They claim that they will allow you to keep the rights to your patent since you are the inventor, and that they will help you promote your invention and your commercial product, and so on. In fact, this cannot be trusted. What guarantees that my invention will not be stolen?

How could you chase after this website that promised to help you, if they are situated in countries far away from you?

You can never rely on what you see on the internet with a matter like this until you yourself visit their establishment and confirm that they are an official, renowned company, with rules and regulations that protect your rights as an inventor. An example of protecting your rights with your invention is through a non-disclosure agreement, that assures the inventor that any details of their invention will not be leaked by the company that they are making a deal with.

Product Specifications and Measurements

The specifications and measurements are a document that determines the specifications and approved procedures to guarantee that a product will be up to standards. It is always done in accordance with certain regulations that ensure the product's adherence to matters like safety standards, weight and measurement standardization, and their compliance with locally approved industrial specifications. The document is considered a fundamental pillar of industries, trade, technology and scientific research. There can be no advanced manufacturing or a fair trade without unified, precise measurements.

The authorities of all countries put a list of regulations that all companies and factories must adhere to in their production of different products. Any product that does not comply with the system of specifications and measurements will not be able to get into the market. Authorities also have laboratories to measure products and confirm their compliance.

Footnote: The chosen system of weights in the Arabian Gulf is kilograms, whereas in the rest of the world it is lbs. Will it be possible to market a scale that uses lbs. to measure weight in the Arabian Gulf?

Consult the authorities responsible for the specifications and measurements in your country to confirm the rules and regulations that you must adhere to while manufacturing your product. These authorities may sometimes be a part of the ministry of trade, or it could be a separate entity. They are usually connected to regional and international organizations, and this is in order to link the local standards to international ones to achieve exchange demand in international trade. For more information, you can visit the following websites:

https://www.gso.org.sa/en/	GCC Standardization Organization
https://www.saso.gov.sa/en/pages/default.aspx	Saudi Standards,Metrology, and Quality Organization
www.iso.org/iso/home.html	International Organization for Standardization

Quality Assurance

Every establishment must have a quality assurance department. This system aims to achieve production without flaws in administrative and technical aspects. This is done by routinely observing work processes and employee performance, and ensuring that work performance matches the specifications and measurements required at the establishment.

Failure of the quality assurance department in its mission has serious consequences on the establishment. This will have a visible effect on the productive capacity of the company and the quality of the product, and therefore the reputation of the company in the market. For example, General Motors, a multinational automotive manufacturing company, have recalled a total of 41,000 of their cars that were being sold in America for fixing in 2012. This happened because of faulty manufacturing that made these cars leak from their fuel pumps, which increased the chance of a fire happening.

Source: (http://www.albiladdaily.com/news.php?action=show&id=110508)

Additionally, in October of 2012, the Japanese Toyota Motor Corporation launched a gigantic campaign to recall 7.43 million cars worldwide from the market because of a defect in the button that opens the car windows. This was the largest scale campaign of its kind since 1996 when the American Ford Motor Company also recalled a big quantity of cars.
Toyota had previously announced in August 2012 that they had recalled some models of cars because of a defect in the suspension in those cars.
Source: (www.bbc.co.uk/arabic/business/2012/10/121010_toyota_millions_cars_recall.shtml)

Another defect in the gas pedal was found in over 7 million cars in the United states as shown in figure (5-1). The pedal would get stuck down on the floor of the car while driving because of its overly low placement.

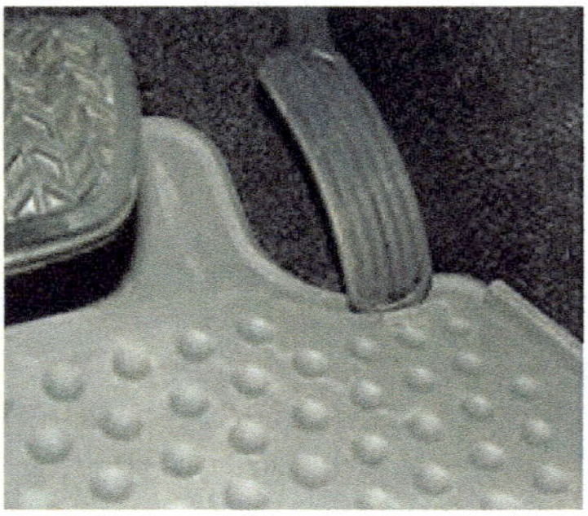

Figure (5-1) Gas pedal defect
(Source: www.mibz.com/28572-us-20-million-cars-affected-by-recalls-in-2010.html**)**

All these losses would not have occurred if there was no negligence in the quality assurance department that eventually led to their failure in proper manufacturing observation and confirming the quality of the different parts, etc.

Product Instruction Manual

Every product that gets manufactured and shipped out to be sold in markets must have a manual attached to it that contains information on how to properly use the product. It should contain information such as how to assemble the device, how to turn it on, where to use it, etc. As for products like medicine, it should list the dosage required depending on the severity of the illness or the age of the patient, etc.

One important thing that should be mentioned is warning the consumer of incorrect usage of the product. Incorrect use could lead to damage to the consumer which will reflect negatively on the reputation of the producing company, which is why it is essential to have instructions on what to avoid doing with the product. This will protect the company from being sued in the event of a consumer having not read the instructions and proceeded to damage themselves.
Figure (5-2) demonstrates examples of these warnings. These warnings are written inside the manual, or attached to the product itself or on its outer packaging.

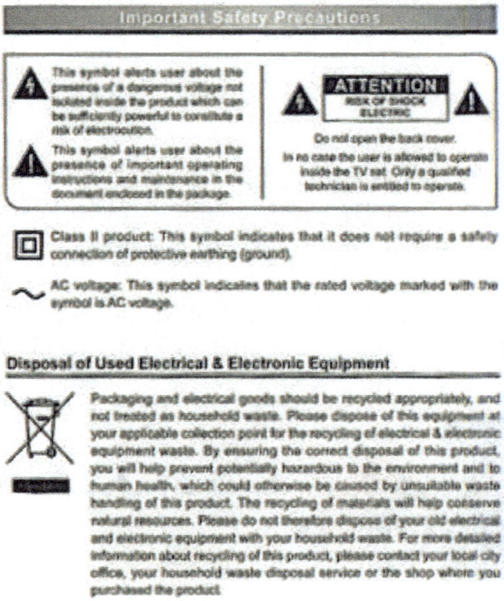
Important Safety Precautions

This symbol alerts user about the presence of a dangerous voltage not isolated inside the product which can be sufficiently powerful to constitute a risk of electrocution.

This symbol alerts user about the presence of important operating instructions and maintenance in the document enclosed in the package.

ATTENTION
RISK OF SHOCK ELECTRIC

Do not open the back cover.
In no case the user is allowed to operate inside the TV set. Only a qualified technician is entitled to operate.

Class II product: This symbol indicates that it does not require a safety connection of protective earthing (ground).

AC voltage: This symbol indicates that the rated voltage marked with the symbol is AC voltage.

Disposal of Used Electrical & Electronic Equipment

Packaging and electrical goods should be recycled appropriately, and not treated as household waste. Please dispose of this equipment at your applicable collection point for the recycling of electrical & electronic equipment waste. By ensuring the correct disposal of this product, you will help prevent potentially hazardous to the environment and to human health, which could otherwise be caused by unsuitable waste handling of this product. The recycling of materials will help conserve natural resources. Please do not therefore dispose of your old electrical and electronic equipment with your household waste. For more detailed information about recycling of this product, please contact your local city office, your household waste disposal service or the shop where you purchased the product.

Figure (5-2) Examples of warnings for consumers (source: Hisense LED TV Set up guide)

Business Administration

As we have previously discussed, establishing and managing your own company requires knowledge in several fields such as the legal procedures of your country and managing internal company affairs. We will talk about a few crucial points required to manage a small company.

Required Licenses

The first step is to work on the required licenses from the relevant authorities. This includes the license to do industrial business, the license to do commercial trading, a municipality license, etc. Consult the relevant authorities in your country to know more about this topic.

Invention Registration

The product, the invention that you will manufacture and sell must be registered in the ministry of trade and adheres to the official specifications and measurements of your country. This will guarantee legal protection for you in the market against any party that tries to imitate your product.

Trademark

A trademark is an icon, whether it is a picture or a letter, or anything else. It is used to distinguish your company to consumers. Every company has their own trademark, and intellectual property laws prohibit anyone else from imitating this trademark.

You can file a trademark registration form with the relevant authorities in your country. You can see examples of trademarks in figure (5-3).

Figure (5-3) Examples of company trademarks

Capital

Funding is an important aspect. Without capital, you cannot establish a company. If you are unable to fund yourself, there are companies that assist inventors in the creation of their inventions by giving them soft loans that make them very little profit or even without profit at all. This is to encourage projects that may be successful and have a role in aiding the economy of the country. These funding companies also provide libraries and workshops for a reasonable fee for creators. The countries of GCC have a hand in funding and supporting small projects by establishing organizations that specialize in this field. If you are interested in these efforts, look at Chapter 6.

Calculating Costs and Profits

There is a lot to calculate when it comes to establishing a factory or a company and all they need to get started. This includes tools, human resources, the cost of monthly operating, etc. Based on this, you will get an idea of how much the minimum cost is for the company's monthly production, as well as determining the price of the product in a way where the monthly income for sales exceeds the amount needed for monthly operating by a reasonable margin.

Example: Let us assume that the monthly operating fee for a factory is 3000 BD, and that the product's price is 30 BD. Dividing the monthly operating costs by the product price gives us the product quantity that must be manufactured. Keep in mind that the costs include the consumption of assets in addition to direct costs and indirect costs, such as electricity, rent and salaries.

Quantity = Monthly operating fee / Product price

= 3000 BD a month / 30 BD

= 100 units

This means that producing one hundred units a month for 30 BD each will cover the expenses of the factory's monthly operating.

But What About Profits?

In order to gain profit from production, sales must exceed the monthly operating costs of 3000 BD. This can be done by increasing the production quantity a month, or raising the price of the product. Both options have opportunities and threats, and you can make your choice depending on the situation of the market.

To guarantee profit, try applying the following equation:

$$\text{(Maximum production output/2)} > \text{Operating costs}$$

This means that the expenses of factory operating must be less than half of the maximum production output. This can be calculated by the day, by the week or monthly.

<u>Example</u>: The maximum production output for the factory is 300 units a month with a price of 30 BD for each unit, totaling to 9000 BD a month.

$$2/9000 = 4500 \text{ BD}$$

Meaning that operating costs must be less than 4500 BD.

Dividing Production Costs

Production costs of any product are divided into three parts:
Expenses of raw materials used in production
Expenses of human resources that work in production
General expenses for the factory such as electricity, water, manager salaries, etc.
Asset expenses, this includes the machines and equipment that are used by the company and are worn out over the time.

Production Costs Stages

The stages of production costs are also divided into three clear parts in the following figure(5-4).

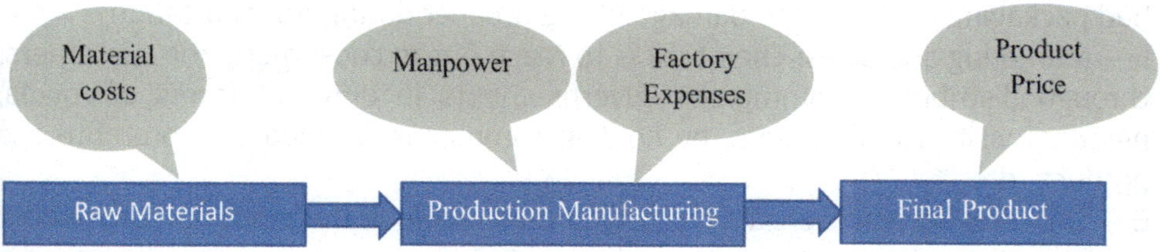

Figure (5-4) Production costs stages

Human Resources

Determining the number of workers and their specializations is an important task, and it depends on the work done within the factory. It must be carefully calculated, as hiring an excess of workers means more expenses that are not coming back, and likewise, hiring too little workers will lead to a shortage in production.

We can organize the workers depending on the following tasks:

1 – Purchases: This gives them the task of buying the required parts and raw materials necessary, either locally or by importing them, to manufacture the new product, in addition to any other company necessities.

2 – Assembly line: Their task will be to assemble different parts of the product in a particular sequence, where each worker would put on a specific part of the product. There can also be a disassembly line in addition to the assembly line to prepare parts. This, however, all ultimately comes down to the type of the product, the ease of the process in addition to expenses.

3 – Quality assurance: As previously mentioned, this is an essential part of any factory, even in commercial companies. It entails observing the production and maintaining the quality by putting criteria that must be followed by all workers. This also entail inspecting the product after manufacturing and assembly to make certain there are no flaws. The lack of proper quality assurance leads to big undesired losses. For example, discovering a flaw in several amounts of the product at the time of inspection means loss of money, time and effort. Loss of time and effort are further indirect losses of money. After inspection is finished, the quality assurance worker puts a special sticker on the product as an inspection certification, in addition to the model number and serial number.

4 – Packaging: After inspection of the product and attaching stickers, comes the role of packaging the product and placing it in its designated box to prepare it for shipment into the market. It also entails attaching the product information on the box if it had not already been previously printed on the box. Quality assurance has a role in determining the suitable packaging and the box for the product as well, as bad packaging could lead to damage of the product during transportation.

5 – Marketing and sales: Their task is to reach out to consumers, the customers, through distributing commercial advertisements in different forms of media, personal visits to customers, by holding commercial seminars to promote the product, etc.

6 – Delivery: No commercial company should overlook the need of a driver that carries out the delivery tasks. Being the owner of the company, you may need a way to deliver products to customers or market distributors.

7 – Finance and accounting: They play an essential task in recording financial matters such as expenses, earnings, profit, loss, etc.

8 – Research and development: Given that the company produces their own unique product, meaning they have not produced any product adopted from another company, then it is absolutely important to know feedback from consumers. This is to further develop the product to be more suitable, as we have explained in Chapter 3 about the feasibility study. If a product stays the way it is without discovering potential flaws and working on enhancing it, it will eventually become outdated. Therefore, research and development have a central role in the longevity of the product and therefore the company as well.

9 – Customer service: Selling your product is often not enough, because the relationship between the consumer and the company continues even after selling the product. The customer may need a service after purchase, such as maintenance, buying spare parts, enhancing the product, etc. Excellence in customer service will greatly build your company's reputation and maintain its position in the market.

10 – Personnel affairs: Naturally, there must be someone who follows-up on workers in the company, starting from recruitment, studying their qualifications, determining their pay, rewards, training, etc.

This is a summary of what the human resources a small factory may need. Keep in mind, that it is possible to assign more than one of those tasks to one person at the beginning. In the future, however, with more work and company expansion, there will most definitely be a need for more workers.

If you have little experience with project management, then it is advisable to join one of the training courses in entrepreneurship. These courses are offered by governmental organizations in some countries in exchange for a nominal fee with the goal of encouraging and assisting creators in establishing their commercial ideas.

Keep in mind that any new commercial work will need diligent work and patience, especially in its first stages until the foundation is strengthened. The stage of establishment is the most important stage. One of the success factors of any company is good worker management and building a sense of trust between the workers themselves, as well as between the worker and the owner.

A successful owner considers their employee to be a partner at work, giving them the respect they deserve, and listens to their feedback on how to improve the work of the company. Unlike unsuccessful owners who always think of their employee as a tool to order around wherever they want, and they always think they know more than anyone else, and do not listen to any suggestions of their employees. This mentality that some owners may have leads to irritation from workers, which, in turn, affects their production, and may lead to the demise of the company.

Part Two: Marketing

Why are some products marketed and sold quickly while others fail in doing so? According to statistics, reasons of product failure come down to:
18% Financial reasons (Low funding)
32% Technical reasons (Technicality, outdated equipment)
50% Marketing reasons (Not studying the market, not knowing the market demands, not studying the benefits, weak advertisements, mistakes in expenses calculation, questionable prices.)
Source: (http://www.abahe.co.uk)
This statistic shows the importance of marketing in the success or failure of selling products. A particular product may be sold in the market more than others, not because of its features but because of the powerful marketing.

From another perspective, do people always know what they want?
Can all people recognize their needs?

You may be able to know what a person needs, whether it is a specific tool or a system that makes their work more convenient before they even realize it.
Many people are content with doing their work traditionally without thinking of ways to develop new methods that may facilitate their work. This is where your role comes in as a diligent creator, to find out the hidden needs and turn them into revenue.
The extent of your success in marketing, and your sales of the invention depends on your ability to convince others of the importance of your invention. Your ability to convince others depends on the importance of the invention to its user. If the invention is a perfect solution for a problem, then it is easy to convince a consumer to purchase it. Then comes the cost and the performance efficiency, and therefore it is important to study the invention's feasibility before manufacturing it, as we have previously discussed in Chapter 3, to estimate and learn the size of the market, meaning, the number of consumers for the product in a specific geographical location. The more the consumers, the more markets there will be for the product.

Persuading Customers

We can summarize the points that will help you persuade a customer to purchase the new product with the following:
Does the new or enhanced product have a positive effect on the environment more than others? In other words, does it produce less pollution?
Does it cost less than its predecessor while maintaining the same efficiency?
Does it consume more power to perform the same level of efficiency?
Does this product have an estimated lifespan?
Will this product be compatible with other, similar products? For example, the product may be an electronic device that connects to a personal computer. Will it be able to connect to all kinds of computers without issue? Or can it only connect to specific types of computers?
If we think back to Chapter 3, where we previously discussed the feasibility study, we mentioned an example with ZIP Drive, and how it could not last because it was not compatible with all computers, and its data transfer speed with some computers was slow, which made it impractical.
Does the new product have any extra features relating to safety?
Does the new product save time and effort for its user? For example, remote control systems have offered plenty of convenience for their users, such as with garage doors, TV, etc. Moreover, the ability to connect to the internet using mobile phones, conducting business transactions, booking flights and more without the need to be present in an office.
Saving time and effort means saving more money. The easier it is to perform multiple tasks seamlessly, the more money it will save on its employer. For instance, the bar code reader system offered a tremendous amount of time and effort efficiency on big stores. This system also accelerates the process of calculating the commodities sold while providing precise information to the storage manager about products that have run out all the while they are sitting in their office. They can then proceed to refill their stock without the need to send out inspectors.

Is the new product easy to use and does not require reading an instruction manual in detail? Many of us do not read manuals or instruction booklets that are attached to a product we purchase out of laziness. This is an undesirable behavior, because any product, especially machinery and electronic equipment have specific rules when it comes to turning them on and using them, and so it is a must to read the tips and instructions attached to know the best way to use the product. With that said, however, the easier it is to use a product without the need to read much, the more it will help encourage a costumer to buy the product.
Think about how your invention could be a reason in evolution of life, and try to specify its benefits in detail. Write a summary about the new or enhanced product detailing the goals of the product, its importance, specifications all to try and persuade customers.

Footnote: A successful marketer knows how to attract customers using the power of words. Suitable words and shorter sentences.

One of the tasks that marketers from companies is learning the demands of the market through collecting information from consumers about their needs of specific ways to make their work easier. People may sometimes need a product without realizing it, and it is a marketer's job to know a consumer's satisfaction of an existing product with the aim of enhancing it, and then relaying all the information to the engineers and technicians at the company that will do their role in meeting the demands of the market.
To reduce efforts in marketing, there must be a focus on those who benefit from the new invention. Let us not forget that marketing also costs money, which means you cannot afford to waste time and money in marketing a product in an area with a low number of consumers or maybe even none. There may be parties that are promoting a product that does the same task as your invention, this is where competition will come into play, and you will have to work harder to convince a customer. So, try to find a market with the least competition. The earlier you get into that market, the better the chance for you to control it and make competition difficult for others. Gaining the trust of a customer will also need effort by providing the best services for a high-quality product with a suitable price and decent service after the purchase.

Footnote: Some car exhibitions allow their customers to test the new car to gain their trust.

One of the strategies that some companies follow in marketing their new product, is handing out free samples to their customers with the aim of testing it, so that they convince them of the importance of the product. In some big commercial stores, we sometimes see a small stand with an employee inviting customers to try new foods such as cheese, sauces, etc. Many consumers often avoid buying new products so as not to waste money on something that may not suit them, and therefore producing companies follow this strategy; to break that doubt between the consumer and the new product.

This also includes good packaging for the product that will give a positive impression to the consumer and encourage them to buy it. It may even get to a point where the product's elegant packaging is more valuable than the product itself, and this happened to me personally as I was buying an educational product that I was quite fond of. I saw it sitting on the shelves of a big store, with fine packaging, and when I opened it at home, the box only had a few simple pieces that are already available in the surrounding environment, and the price paid for the product was not worth it. This, of course, is wrong, as it is an attempt trying to trick a consumer with the external appearance of a product without the existence of value for the product itself. Avoid doing these methods that may have a negative impact on your company's reputation.

Gaining a Customer's Trust

Footnote: A quote of the late Dr. Ibrahim Elfiky: A successful marketer possesses principles like believing in themselves, their abilities, their company, and their product.

Do not make it a priority to blindly sell a product without thinking of the customer's interest. Always make the customer feel like you are with them, let them know that you are there to help them with their problem. For example, if a customer needed a specific system, like a computer application to help them organize information of books, tell them that you are ready to help them solve that problem by enhancing that system.

In the process of finding solution for customers, always be flexible, open-minded, a partner to the customer while solving their problem. Be close to them, call them after they purchase the product and make sure everything is fine and that they are comfortable with the product. When the customer sees this treatment from you, they will come to trust you.

A customer's satisfaction, without a doubt, is the key to success, development and longevity of any company. It is a part of quality assurance departments in pioneering companies.

Presenting the New Product

How will you present your new product to customers, whether the customers are the users or investors that you require the help of to reach customers? An investor may be an owner of a big store, or a local product distributor, for example.

To be a great presenter for the new product, and to be able to persuade your customer, you must know how to speak, but what do we focus on that will attract a customer's attention? Try to display the many positives of your product and compare it to another product that does the same task, and talk about what makes it stand out in terms of price, ease of use, ease of maintenance, efficiency, and more, without getting into the theoretical part of how the product works and its components, this is not what interests customers unless they ask for it. If that happens, you may try to answer them, but you should still try to limit the information you give away.

Try to give a summary without any excess talk that may bore the customer. Try to be ready to answer any question a customer may ask. You must also be confident while presenting, as you may have to do things like handing out informational leaflets as in figure (5-5), and bring test models for the product to try in front of the customer.

Figure (5-5) New product poster

In this regard, I will show you an article from March 17th of 2014 that I have read and liked, and it goes by the title "Avoid These Five Sentences While Presentation to an Investor" by Robert Carroll. You can find it in the following link: http://ar.wamda.com

Believe it or not, investors are the ones that want to give their money to you, all of it. They will present enormous amounts of money if you have a great work team, a great product and if you studied the market carefully. At the same time, there are many means that make you flop and miss the opportunity of receiving funding. To help you excel in presenting your idea, I have collected a list of five things that you should not say while presenting to an investor:

1 – My partner works part-time

If your idea is worth 100 million dollars, then none of your partners would want to stay in their current job. If you are unable to find a person that devotes their full time to your idea, then is your idea really worth 100 million dollars?

2 – We have no competition

This does not apply to start-up companies. For every product you will find a competitor, even if it is a disposable product. Do not ignore competition because you must go into the market ready. Contrary to popular belief, sharing the details of your company will not harm it, it will actually benefit it. Having competition proves that your idea is great and there is demand for it. If the investor notices that you understand your competition well, they will think you are smart enough to beat them. If you understand your opponent and yourself well, then you can fight a hundred battles without damage.

3 – We need 15,000 dollars to rent an office

If Steve Jobs was able to build Apple starting from a garage, then you are able to build a start-up company from a net café. Always remember that Steve did not have internet connection back then.

4 – We have a clear exit strategy

Some investors may not agree with me, but I support Mark Cuban in his saying: "***Don't start a company unless it's an obsession and something you love. If you have an exit strategy, it's not an obsession.***" (Source: http://www.entrepreneur.com/article/222524) What investors want is to see that you are passionate about your idea and company. They want to work with incredible pioneers ready to do their absolute best to turn their ideas into successful companies. If selling your own company is one of your priorities, then you are not doing it right, and know that money has never been as strong an incentive as passion.

5 – I am somewhat worried

Do not think of presenting to an investor as a very serious matter, try to be as laid back as you can. Wear a pair of jeans and a normal shirt if that makes you feel less tense, and remember that you are not in Wall Street. Try to let go of the formalities and make your presentation more akin to a friendly, pleasant conversation.
There are several things you can do, such as making use of commercial exhibitions that are held to display products, having your product in commercial magazines, making your own website or page on social media like Facebook, Instagram, or Yellow Page, etc.

New Product Summary

Writing a summary about your new product is required when you want to present your product to investors or manufacturers. This summary must contain the most important points without all the boring details.
Try to make a summary of one page, mentioning the background of this new product, its benefits and features. This summary must look inviting for investors and manufacturers, with a stylish design in terms of the writing and the images. Do not forget that that owning a patent or any kind of approval certificate from academic or research parties for your product will increase the product's value and gives it a push.
In the third appendix at the end of this book, we will have an example of a new product summary.

Chapter 6
Patents

We often hear about patents, and how someone got a patent after creating something, and sometimes we hear about conflicts between companies in courts regarding ownership of a patent, and this happened between Apple and Samsung around the invention of smart phones.
What do we mean by patent? What is their purpose? How can one be obtained? What are their requirements? What benefits does it give its owner?
We will answer all these questions in this chapter without going into too much detail.

Defining Patents

A patent is a document that is given by the relevant authorities to a person or a group of people that had a hand in the creation of a new tool, device, material, or component to prove that they were the first ones to do so. A serial number is given as a reference to this patent, and it is also announced in magazines and journals. A patent is applicable to all kinds of devices, tools, machines, treatment processes and the creation of new materials. Things like mathematical rules, physical phenomena, and software are not applicable, and instead are considered intellectual property.

Obtaining a patent entails its owner getting full rights of manufacturing and marketing, and prevents any other person or party from doing so without the permission from the owner of the patent.

Registering a patent may take a lot of time and effort, depending on the party that you go for to apply for a patent and the kind of invention. There are several documents that must be presented to apply for a patent, they must include all the details and explanations about the new invention including conceptual designs. There will also be a need to pay fees to apply, and it may sometimes require legal consultation to continue, and this will have additional costs. Because a patent protects its owner when it comes to manufacturing and marketing, companies tend to register patents and take every necessary step to protect their products in the market and guarantee sales and competition. Therefore, it is not advisable to spend money and effort in registering a patent if there is no profit in it, or in other words, if the product cannot be sold for one reason or another. On the other hand, your invention may be beneficial later and you will get all the credit then, and it will also be great for your CV.

Some companies manufacture new products and push them onto the market without registering a patent, and there are a few reasons for that:

They cannot afford to wait to obtain the patent, so they push the product into the market to sate the needs of consumers for the product. For example, the product may be a commodity for winter such as a heater. This product must be manufactured and pushed into the market by the end of the summer. If they were to wait until obtaining the patent, then the season for that product may already be over.

Expected profits from the product are low, and so they save money by not investing in a patent.

Some companies have the ability to distribute a product in the market rapidly, and by the time competitors realize that and make the same product and distribute it, the first creator would have already swept through the market and competition will be difficult, and this is a strategy.

The more complex an invention or product is, the more guaranteed it is that the design will be difficult to imitate. Some companies, in order to prevent their creation from being stolen, remove numbers and symbols that are written on the electronic parts, or paint over circuits to hide the wiring, and so on.

Patent Registration

There are many organizations that receive documents needed to register a patent, and there are local companies in most countries known as national patent offices, and other are regional like the following:

1 – The Gulf Cooperation Council (GCC) Patent Office: It is a patent office for the Arab States of the Gulf, and it is a regional office with their headquarters being in Riyadh in Saudi Arabia. Their patents give their owner legal protection for the product in all six countries of the council and are active in all of them starting the date of granting the patent, and it is valid for twenty years starting from the date of handing in the request for the patent at the office.

In addition to that, this office encourages scientific and technical research, and invigorates the market in the Gulf States which reflects positively on the economy of the Member States.
For more information on the office, how to apply and the relevant rules, please visit the website of the office at www.gccpo.org.
2 – United States Patent and Trademark Office (USPTO): This I an office for registering patents as well as trademarks in the United States of America. All patents by this office prevent others from manufacturing, selling, and using the invention in the US.
For more information on the office, how to apply and the relevant rules, please visit the website of the office at www.uspto.gov
3 – European Patent Office (EPO): This office was established in 1973 in Munich, Germany. It provides patents, in addition to a service called Espacenet; a service for searching and studying over 70 million patents that were given throughout the world from 1836 to the present day with the largest database in this field. The office also has another excellent service, and that is the translation of patents from any language that is not English, to English.
For more information on the office, how to apply and the relevant rules, please visit the website of the office at www.epo.org

Protection of a Patent Internationally

It is possible to do so through the Patent Cooperation Treaty (PCT). This treaty was signed in 1970, and has been modified since. This treaty ensures legal petition for the invention in accordance with a patent in many of the contracting countries, and they are 186 countries, by requesting an international patent. If you are a citizen or resident of one of the contracting countries, then you can file your request at the local patent office. It is also possible to file your request at the World Intellectual Property Organization (WIPO) in Geneva through personal attendance or through the organization's website. This organization promotes creativity to achieve economic, social, and cultural development for all countries through an international system for intellectual property that is characterized by balance and efficiency.
For more information on the treaty, please visit the website of WIPO at www.wipo.int

Misconceptions

There are some misconceptions surrounding patents, and these are:
1 – You must be a genius scientist to obtain a patent: Obtaining a patent does not need a genius scientist with the highest qualifications and the best academic certificates. If you merely enhance an already existing invention, no matter how simple this enhancement is, as long as no one else has previously done it, then you deserve to get a patent. There are countless inventors through history that have not even finished their studies, and instead made time for their creations that left their mark in history.
2 – An inventor must be an engineer: A person wanting to invent something does not necessarily have to had studied engineering. Your current profession and experience may be enough to invent something and obtain a patent.
3 – You will get an excess of money as soon as you obtain the patent: In some cases, this may be true if the invention is highly essential and can be directly sold to a company that will handle its manufacturing and marketing. In most cases, however, it requires diligent effort on your part to turn an idea and conceptual designs into reality, and this is what we had already discussed in this book. Keep in mind that the time it takes to think of an idea, and then register a patent, is a small fraction of the overall time that you invest in the invention.
4 – A patent is not possible because the invention is too simple: It does not matter whether the idea is simple or complicated as much as whether it is new. There are many simple inventions that became widespread and had a role in changing people's lives for the best. For instance, the telegraph is an invention that had a switch that would send intermittent electric currents to a transmitter that is made of a magnetic coil and a metal magnet piece. This simple communication invention helped the American society be closer to each other in the 19th century.
5 – A patent is not possible for enhancing an already existing product: You can, in fact, obtain a patent for enhancing an invention that was already made, as there is almost no invention that remained the same since its inception. For example, in the 19th and 20th centuries, there were 1673 patents for washing machines. Every one of those patents was an enhancement in the performance of washing machines.

Methods of Patent Searching

Before presenting your patent registration documents to one of the relevant authorities, do a bit of searching by yourself and make sure there is no other similar patent, because this will most likely lead to rejection for your request, and so it is best to make sure first. This allows you to make changes to your inventions that makes it different from what already exists, and therefore guarantee obtaining a patent while saving yourself a lot of time, effort and money.

On the other hand, in some cases, you may not even need to search for any patents and you could go ahead and file your request directly. This applies only if you know that there is nothing like this invention. Examples of that include television and optical computer mice. You will most likely already know whether this invention already exists if the invention happens to fall under your profession or specialization.

Here are some means you can use to search for patents:

1 – Start with a search engine such as Google for anything similar to your invention, and start searching for images because that may be easier. If it already exists as a commercial product, then you will eventually find a picture since companies display pictures for their products on the internet.

2 – Search on the Google Patent Search website, which has the same registered patents in the US office, but using this will be faster.

3 – Search the records of patent offices that we have already mentioned through their websites.

4 – Use an application called PatentHunter. This is an application that can be downloaded from the internet. It has the ability to access databases of the US patent office, as well as the European one. You will be able to search any previously registered patents and download documents related to them.

5 – Search in PATENTSCOPE, owned by WIPO on the following link http://patentscope.wipo.int/search/en/search.jsf

Searching for patents is not an easy process, because there are millions of patents in all the different fields, and you will not find all of them in one place. The process may be like trying to find a needle in a haystack; so, knowing the professional methods of searching all the databases is important.

One of the problems you will face while searching is choosing the most suitable word to search, which is why you may have to use different words as synonyms to your invention. For example, the word "glue" has synonyms such as adhesive, binder, paste and gum. So, you must use as many words as possible that you can think of and make use of a dictionary to learn more synonyms.

You can search using the task that the invention will do instead of the invention name, which may prove useful. For example, if a kind of glue was used only for certain materials, then we can search using the material name, such as wood glue, wood joint or wood adhesive. We may also use the word "furniture", and so on.

Researchers that work at patent offices are professionals with much experience that allows that to make sure whether an invention already exists.

References

Part one: Books and research

1	From Patent to Profit: Secrets & Strategies for Success. Bob DeMatteis 3rd Edition SQUAREONE Publishers
2	From Invention to Innovation: A guide to developing and evaluating your inventive idea. Dr. Sally Caird & Prof. Stephen Potter, Open University, UK. Oct 2010
3	Innovation Process. Andrei Kolodovski for Riso National Laboratory, Denmark. Dec 2006
4	Your Guide to Entrepreneurship Engineer Muhannad Jibril Abudia
5	Muslim Heritage in Our World Professor Salim Al-Hassani Foundation for Science Technology and Civilisation
6	Plug-In Electric Vehicle Handbook for Consumers. U.S. Department of Energy

Part two: Websites

http://www.saso.org.sa/Arabic/Pages/Default.aspx	Saudi Standards, Metrology and Quality Organization
//en.wikipedia.org	Wikipedia
www.businessballs.com	Business Balls
www.gccpo.org	Patent Office at the Secretariat General for Cooperation Council
www.uspto.gov	United States Patent and Trademark Office
www.wipo.int	World Intellectual Property Organization
www.epo.org	European Patent Office
innovationcentre.ca/	Canadian Innovation Centre
www.furnituretest.com/	Advanced Furniture Testing
www.about.com/	
www.youmotivation.com/	YOU MOTIVATION
www.abahe.co.uk/	Arab British Academy for Higher Education
www.gso.org.sa/	GCC standardization Organization
www.moic.gov.bh/Ar/Pages/Home.aspx	Ministry of Industry and Commerce – Kingdom of Bahrain
http://www.arceit.org/	Arab Regional Center for Entrepreneurship and Investment Training

Appendix 1

SWOT Analysis Template

State what you are assessing here ____

(This particular example is for a new business opportunity. Many criteria can apply to more than one quadrant. Identify criteria appropriate to your own SWOT situation.)

Criteria examples	Strengths	Weaknesses	Criteria examples
Advantages of proposition? Capabilities? Competitive advantages? USP's (unique selling points)? Resources, Assets, People? Experience, knowledge, data? Financial reserves, likely returns? Marketing - reach, distribution, awareness? Innovative aspects? Location and geographical? Price, value, quality? Accreditations, qualifications, certifications? Processes, systems, IT, communications? Cultural, attitudinal, behavioural? Management cover, succession? Philosophy and values?			Disadvantages of proposition? Gaps in capabilities? Lack of competitive strength? Reputation, presence and reach? Financials? Own known vulnerabilities? Timescales, deadlines and pressures? Cashflow, start-up cash-drain? Continuity, supply chain robustness? Effects on core activities, distraction? Reliability of data, plan predictability? Morale, commitment, leadership? Accreditations, etc.? Processes and systems, etc.? Management cover, succession?
Criteria examples	**Opportunities**	**Threats**	**Criteria examples**
Market developments? Competitors' vulnerabilities? Industry or lifestyle trends? Technology development and innovation? Global influences? New markets, vertical, horizontal? Niche target markets? Geographical, export, import? New USP's? Tactics: eg, surprise, major contracts? Business and product development? Information and research? Partnerships, agencies, distribution? Volumes, production, economies? Seasonal, weather, fashion influences?			Political effects? Legislative effects? Environmental effects? IT developments? Competitor intentions - various? Market demand? New technologies, services, ideas? Vital contracts and partners? Sustaining internal capabilities? Obstacles faced? Insurmountable weaknesses? Loss of key staff? Sustainable financial backing? Economy - home, abroad? Seasonality, weather effects?

Appendix 2
Searching for Information on the Internet

Efficient Search Methods

Seeing as the internet is an ever-expanding world and it constantly grows, you will need skills that will help you reach the information you need. There are many search engines, Google being the most popular one. Let us look at how to use the search tools available to facilitate narrowing down search results.

Search Categories

When you type a word or the name of something that you are looking for, the engine will show you thousands of results across tens of pages. After the results appear, notice that at the top of the page, there are few categories. All, Images for pictures, Videos for video clips, and the last category has more options in it as figure (1) shows.

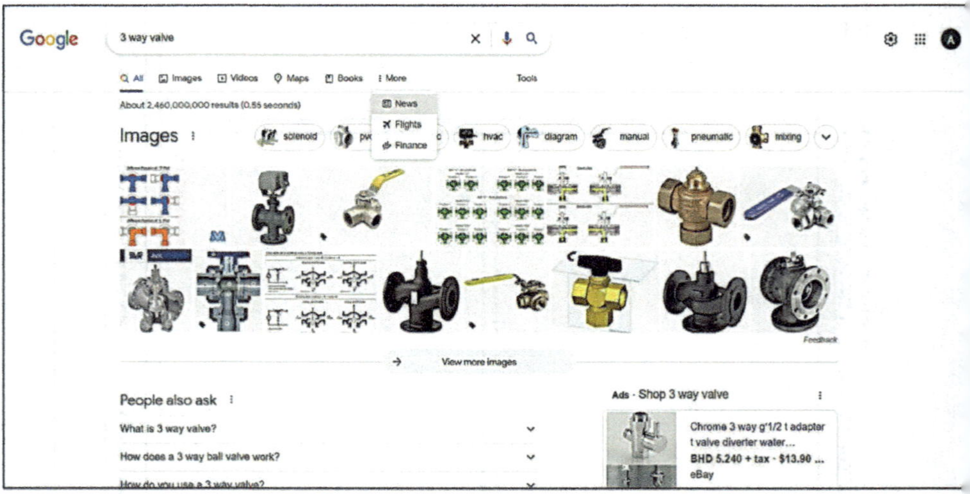

Figure (1)

In addition to that, there are "Search tools". If you click on that, there will be two new categories as figure (2) shows. Searching using these categories will allow you to get varied results for a word or sentence that you are looking for, because the results will differ depending on which category you select.

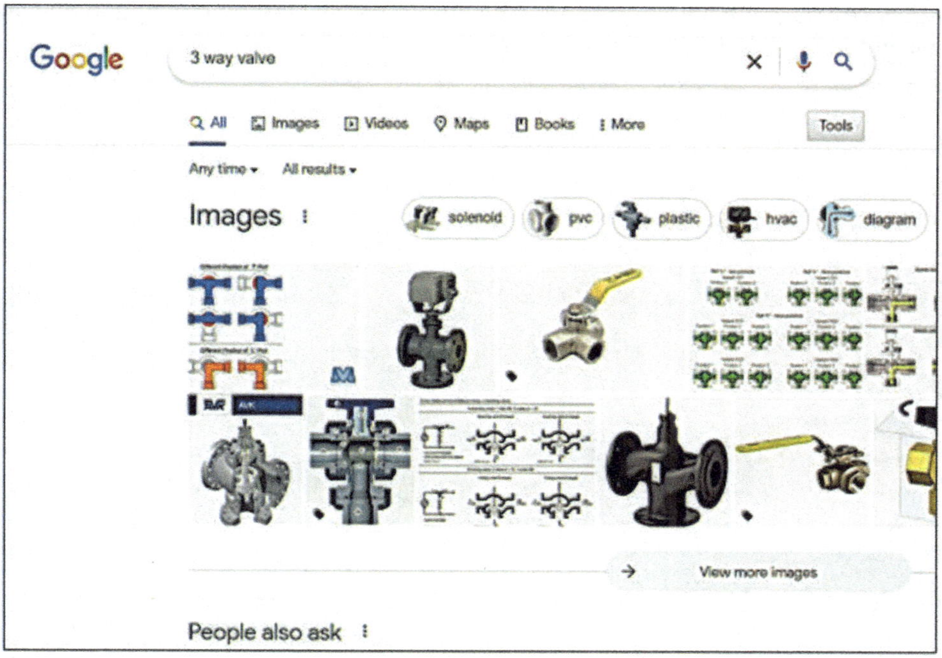

Figure (2)

Advanced Search

Advanced search can be found at the bottom of the page. You will find an option for advanced search under "settings" menu as figure (3) shows.

Figure (3)

Advanced search allows you to search in more detail. Click on this option and you will get a page that looks like figure (4).

Figure (4)

By choosing this option, you can specify the way of writing the name or sentence that you are looking for, and this will help you narrow down the results. Moreover, you can also search by file type. PDF files are usually for reports and researches, and PPT files are usually for lectures, and there are other types of files. You can also specify the language that you want the results to appear in. There are plenty of other options that will help you reduce the number of results to get the desired ones.

Searching Company Websites

There are websites that sell products of all kinds. Electronic products, electrical products, chemical products, stationery, home supplies, kitchen products, clothes, etc. You can search in these websites for the availability of a product that is similar to your invention, and if there is something like it, then you can work on making changes to your creation so that it can be something new.

Examples of those websites are the following:

1 – Ali Baba, a Chinese website for buying and selling on the internet.
www.alibaba.com
2 – Amazon, an American website that started selling books on the internet, and later became a website that sells different products.
http://www.amazon.com
3 – European Patent Office, it has a patent database of over 90 countries around the world, so you can search for any invention that may be similar to yours.
www.epo.org

An Example of a New Product

YCN ***Your Company Name*** *company address, city, phone and fax numbers, website*

New Product Summary—M2K Square Bottom Bag

Product Background

Paper bags have been the most popular style of bag to carry fast food and food service items in high-volume outlets in North America and much of Europe and Australia. Paper bags are the packaging of choice due primarily to their ability to square out and stand up.

In some fast food outlets thin-gauged plastic T-shirt bags using space-consuming racks to support the bags for loading are used. Generally speaking, T-shirt bags do not have the aesthetics sought by the fast food trade, are difficult to load, and don't stand up after being packed.

Since plastic T-shirt bag racks require too much space, have poor aesthetics, are not cost effective in smaller sizes, and do not cooperate with condensation problems, they are not a suitable paper bag replacement. Thus, paper bags dominate the fast food industry until a suitable alternative has been developed.

Product Summary

M2K bags combine the best attributes of both paper and plastic into one single product. They can be loaded faster than paper and stand up just like paper after loaded. Using M2K bags makes it a simple task to check order accuracy and speed customers through the drive-through lanes.

M2K bags have also successfully eliminated the condensation issue associated with plastic. The proprietary "Automatic Ventilating System" (AVS) automatically opens and vents condensation after the bags are loaded.

The proprietary hook dispensing system eliminates the need for space consuming racks. The hooks allow different bag sizes to be placed neatly and conveniently along the side of a counter top. The bags can be quickly loaded, are self-opening when dispensed, and stand up afterward.

Product Benefits

There are many benefits for supermarkets. Here is just a partial list:

- ❒ Lower overall bag costs
- ❒ Lower labor costs
- ❒ Improved productivity
- ❒ Convenient, easy to use
- ❒ Little training required
- ❒ No space-consuming racks required
- ❒ Bags automatically ventilate allowing condensation to escape
- ❒ Per unit cost is 20% less than paper

Patent Status

Seven U.S. patents are pending in the U.S. with international rights reserved.

The Opportunity

A limited number of licensees in the U.S. and internationally are sought.

Source: From Patent to Profit, Bob Dematteis

Made in the USA
Monee, IL
03 November 2026

67380834R00069